Universal Evolution

&

Self-realisation

Second Edition

Published by **Tatterdemalion** Blue in 2025

(First Published by **Tatterdemalion** Blue in 2021)

Words © Mark Callaghan 2025

universalevolution.co.uk

A CIP catalogue record for this book
is available from the British Library

Cover design
by Mark Callaghan and **Tatterdemalion** Blue

ISBN 978-1-915123-15-2

Tatterdemalion Blue
74 Maxwell Place
Stirling
FK8 1JU

www.tatterdemalionblue.com

Universal Evolution

&

Self-realisation

Mark Callaghan

The author was born and resides
in Scotland, as of 1950.

Dedication

*This work is dedicated to Sri Mataji Nirmala Devi Srivastava
and to all those who seek truth, before and beyond belief.*

Contents

Preface .... *i*

Introduction .... *iii*

Part 1

Chapter 1
The Idea of Evolution *1*

Chapter 2
Universal Evolution *17*

Chapter 3
Domains and Hierarchy *33*

Chapter 4
Natural Selection and Ecology *38*

Chapter 5
Consciousness *70*

Chapter 6
Intelligence *118*

Part 2

Chapter 7
The Observer *135*

Chapter 8
The Domain of Culture *161*

Chapter 9
Identity and Self-esteem *189*

Chapter 10
Self-awareness and the Nervous System *211*

Chapter 11
Self and Self-realisation *238*

Chapter 12
Dirty Hands *259*

Postscript *263*

Preface

This book is really just a long essay about evolution, in which I discuss some ideas and perceptions around the subject. Some of these are already well known, some are my own, gathered over many years. Generally, people tend to think of evolution as an obscure and highly specialist subject best left to experts to consider and figure out. I used to feel that way myself, but I now have a very different, profoundly interested but relaxed attitude towards it. Evolution is about you and me, it's about the world we inhabit and belong to. It's not just a specialist's subject. It's for anyone and everyone to consider (or not if they prefer) and add to the sense of being part of a greater whole.

Generally, the word evolution is used in the context of organic development, life that is, but it has come to be used now in other contexts also, particularly that of the development of stars, galaxies, and the elements of which these are composed. Close observation suggests that these apparently very different forms of evolution may not be so different after all. No distinct dividing line has been found to exist between animate and inanimate nature. There appears rather, to be an unbroken line of continuity from events and processes at the molecular level to those involved in the processes of life. This gives rise to the perception that at least some of the principles of evolution that can be observed in life are already present and represented in inanimate nature. The purpose of this work is to inquire into what these common principles may be and offer an inclusive perspective of evolution from the beginning of the known universe to the emergence of human consciousness...

In these terms, evolution is a phenomenon that encompasses everything, viz the title, Universal Evolution.

Introduction

Evolution is an idea that has been a focus of great interest, discussion, and controversy ever since it emerged fully into the attention of Western society, courtesy of Charles Darwin (and equally, but less well known, Alfred Russell Wallace) around the middle of the 19th century. Prior to then it had been assumed in Western societies, indeed specified by religious authorities, that all species had been fixed in form since the beginning of time, and so this idea generated intense controversy. But at the time of Darwin and Wallace's observations and deductions, the scope of Western scientific inquiry had already opened up the recognition that, for one reason or another, species were not fixed for all time. This was inspired particularly by evidence garnered through growing understanding of geology, and interest in fossils which were being collected worldwide, but also because the growth of empirical science generally was beginning to dissolve old and ingrained ways of thinking based on theology and doctrine rather than evidence. So, the idea of gradual change as such wasn't entirely new, and various ideas about evolution were already being explored. What was new was the observation and detailed description of some significant underlying principles, and the scale of change and time involved, all of which was particularly controversial then, and still is for many to this day. This took place in the context of the approaches and methods of Western science at that time, which were rooted in already well-established principles developed by, in particular, Galileo and Newton as well as many others before and since. (A major contribution to their insights had been the invention and development of the microscope and telescope, from the late 16th century on into the 17th, which opened previously unknown spheres of exploration). At the time of Darwin therefore, his observations were interpreted according to notions of the workings of nature prevailing at that time in Western science. However, many of the understandings of science then have long since been superseded by further discoveries and insights, but still bear a considerable influence on our ways of thinking. One of the challenges here is to explore some of these considerations, and in the course of doing that I will necessarily be asking the reader to examine their own understanding of some basic concepts which we may use every day and take for granted without considering too closely.

The kind of illumination science can provide is however, limited to the analysis and description of 'how' things work and are related, the 'mechanics' of reality you could say. Also, one of the drawbacks of the normal scientific approach to exploration and inquiry is that the more highly specialised someone's field of knowledge becomes, the more inaccessible and refined it tends to appear to others, and correspondingly isolated. This approach to knowledge has proved extremely effective in providing the ability to take control of and manipulate many natural processes but it doesn't and can't, provide answers to the broader range of decisions and questions that all human beings must deal with every day and at all scales - particularly where ethics and morality are concerned. These areas also depend on the evaluation of evidence, but of a much less tangible kind, and extend into many questions about existence that all human beings must ask themselves at one time or another, regardless of their background or education. Many questions faced by the human mind can't be expressed in words and are beyond the scope of any form of science to answer beyond those terms of 'mechanics'. Such questions concern the core of conscious experience; 'Who am I?', 'Why do I exist?', 'What is existence...?'. Inquiry and descriptions of this kind appear in terms very different to those of Darwinian theory, but don't necessarily conflict with it. An implicit recognition of evolution throughout nature in spiritual terms, i.e., in terms of 'conscious existence' has been a fundamental feature of much Eastern philosophy and religion for millennia. Part of my effort here is to present ideas about evolution that may help make some connections between these two apparently very different ways of thinking. Part 1 of this writing attends primarily to outlining developments in current Western thinking about evolution, the second part extends this to considerations of human culture, and the relationship between the individual and society. In that section, amongst other things, I discuss some close links between modern Western-based knowledge and that of Eastern philosophic inquiry. The knowledge and success that science has brought to human endeavour has - with good reason - brought many people in Western societies to feel that the scientific approach to understanding the world provides in many ways a more valid view of the nature of existence than older, theological, or metaphysical kinds of interpretations which it has to a large extent displaced. At the same time the essentially technical descriptions and insights of natural processes and explanations in terms of 'how' offered by the sciences leave most people outside of those specialities feeling proportionally ignorant and still just as bemused by the profound mystery of existence as ever before. In some ways, to both those within and those out-with it, science represents a new belief system, with its own protocols and mysteries of knowledge, and a new form of 'priesthood' of authority. There is always some risk involved in creating new frameworks while breaking old ones.

Nevertheless, I have great respect for science, with its emphasis on observing and testing evidence (I would just add the caveat though, that science under an ethos of political interests and commercial imperatives by no means always lives up to its own high ideals). I emphasise though that I'm not a scientist or academic, I'm an everyday human with no special training, just fascination and curiosity - though I'll admit to being widely, if lightly, read science-wise. This has certain advantages. One of these is that while I have great respect for the approach to evidence and truth that science represents and take pleasure in following its explorations, I don't have to satisfy its very challenging requirements - I have more freedom perhaps to speculate than a scientist or academic. What I have to offer here is an informal and personal collection of ideas, and I don't attempt to provide reference details for most of the information that I draw on. Some of my own ideas and observations which I have been considering over many years, are included, and I have also discovered a great deal in the course of writing. In this respect I have the great good fortune to be doing so in an age in which, like me, most readers will be able to check details and explore further online with a few clicks of a mouse or keyboard. Still, even speculation must satisfy the basic requirements of reason and logic, otherwise it is no more than fantasy, and I try to keep these as my guides.

My qualifications, therefore, are simply those of being alive and possessing a brain provided by the efforts of several billion years of evolution, which I take to be qualification enough for the holding and expression of ideas and opinions at least... This work should be regarded as simply one person's compilation of perceptions for what they may be worth as a contribution to discussion. For that reason, what I offer here I present in the first person. Nothing in the ideas here should be taken as strictly factual - it is a work of exploration and curiosity (and reason, or so I like to think) - so don't be fooled by my style of writing, this is really a work of fiction, or let's say, of imagination, rather than science. Having said that, it isn't possible to tackle the idea of evolution without dipping into several areas of science that have provided us with important insights that contribute to our understanding, and I do so freely where appropriate. Many of these scientific insights provide 'vertebrae' to the structure of ideas I offer here. Part of the challenge for me in presenting this is that it's necessary to try to balance between being too simplistic for the more experienced scientific reader and too complex for the everyday reader with little familiarity with science. With that in mind I try to focus on issues central to the main thread, with a minimum of discussion of scientific considerations. Many of these are nevertheless unavoidable and should hopefully be interesting. I should in fairness forewarn you the reader, that in any case in one way or another, this is likely to be a challenging read.

Practical questions concerning the origins of life are being explored extensively in the sciences, from many different directions, and to depths that are far beyond the reach of my knowledge. My intention is not so much to provide any new detailed definitions or answers but rather, to offer as best I can, a holistic view of evolution with some personal observations. In one way or another, this may open up to view for the reader certain insights, most of which already touch the edge of most people's attention from time to time and may encourage further exploration. I should emphasise that the use of all terms in this work is based on my own understanding and shouldn't be presumed to be scientifically accurate. Occasionally I also use terms which seem suitably descriptive for my purposes but won't necessarily appear in any dictionary in quite the way I've employed them... I do try to explain what I mean, in context, as clearly as I can as I proceed.

Regarding established basic principles of evolutionary theory, as far as I can, I present outlines of these as integral to the ideas here. These days, questions surrounding the evolution of life engage the attention of people from many different fields of interest, scientific and non-scientific, which makes it a particularly multi-disciplinary area of inquiry. Accordingly, there are many competing ideas, but within scientific fields there is general agreement that, through processes of gradual transition, life has arisen spontaneously from inanimate nature, and inquiries focus for the most part on mechanisms underlying how this transition may have taken place. The observations and ideas I present here share this orientation but offer a particular way of considering the meaning of evolution that is much broader in scope than that usually taken of its expressions in organic life alone. Let's say that the gene, while still hugely important, takes a somewhat less prominent position in this perspective than is often the case. I expect that there are many errors of detail in this work, not least because of the limitations of my personal knowledge and understanding, but also because new discoveries and adjustments are being made constantly. Despite such errors though, I believe the main outlines to be sound, as far as they go.

As I was writing, I believed the central concept I was presenting to be my own, and quite original, or so I thought... But by the time I had more or less completed my own main thought processes, I had begun to realise that this idea can't be that original; the line of reasoning is so obvious, someone must have followed it already. In fact, a little exploration made me aware that this subject had already been approached over a hundred years ago by Ukrainian mineralogist, Vladimir Vernadsky, and some decades later by palaeontologist, Pierre Teilhard de Chardin. Vernadsky was a scientist, and apparently atheist, Teilhard was a scientist and Jesuit priest. In both cases the evidence that inspired them, at least in the first instance, was clearly that of practical observation not metaphysics, where their views diverged considerably. I have recently come across outlines of their ideas (deliberately, I still haven't read any of their actual work, just available outlines) and there is clearly a resonance in this material, with both of those. It's not an entirely new idea then, but the context of understanding has moved on in many ways in the intervening decades. For some readers these ideas will be completely new, for others, hopefully there may be some new food for thought.

Having taxied into position, before take-off I would like to make a point here that in my personal view, when examined closely the phenomenon of evolution is indisputable by any reasonable observer, although understanding and interpretation of its principles and implications are still very much a matter of great ongoing inquiry and debate. I declare this assumption and won't be engaging in discussion around the validity of the central concept of evolution as this seems to me to be a digression no more relevant or helpful to serious discussion than considering whether the earth is flat or spherical. It's not my purpose here to consider directly for instance, matters such as the difficulties the idea of natural selection creates for literal interpretations of religious views on creation, just to look as best I can, in a general way, at the evidence and information we have about how evolution works and to encourage others to do the same, with an open mind, without prejudice. This is by no means to say that there is only one way of interpreting the evidence. Finally, some of this material will be tough to digest. I would advise the reader not to get bogged down in details. Perhaps skip through it briskly first, then read it again more carefully. It's a sketch, not a great work of art...

Part 1

Chapter 1

The Idea of Evolution

Outline

The idea of evolution initially came to wide public attention with the publication of Charles Darwin's book, 'On the Origin of Species' in 1859. At that time, the term evolution was conceived of and applied to, the field of living organisms alone, and in general, this remains the case. Over the last century however, exploration within the fields of physics and chemistry has made it clear that inanimate nature has also undergone its own form of evolutionary development. This is conceived of in terms very different to those of organic life, nevertheless certain principles of evolution can be seen to apply to both of these areas, animate and inanimate. This naturally creates the perception that rather than being fundamentally different phenomena, these forms of evolution exist on a continuum. In other words, there is one common principle, or set of principles, at the roots of them.

Over this and the next few chapters I will discuss some established perceptions and principles of evolution, along with some perhaps less well-known ideas.

Of course, we perceive living and inanimate objects as being fundamentally different, but when we try to define the difference, it isn't necessarily easy. It has long been apparent, without the need for any deep analysis, that living organisms are composed of material that exists in the environment in an inanimate state. When organisms die, they decompose and return to being

1

inanimate material. From 'earth' we come, to 'earth' we return. It's not so long since it was generally thought that creatures such as flies and mice arise spontaneously from decaying or other inert forms of matter, but with the nature of their species established and fixed according to 'divine design'. In the course of experimentation, inquiry, and observation, largely over the last couple of centuries, it has been confirmed that life does indeed emerge from inanimate material, though not in quite the way that had earlier been conceived... More recently still it has come to be understood that this development began initially over three billion years ago, through many degrees of increasing complexity rather than through some sudden and mysterious transformation. Courtesy of Darwin, Wallace, and others, we have come to recognise and describe many aspects of this process of growing complexity within the sphere of recognisable organic life as we know it, as 'evolution'. Accordingly, we generally associate the term 'evolution' with living processes. Since Darwin's time however, a great deal more has been explored and discovered concerning the workings of nature, both animate and inanimate. Amongst other important developments in our understanding over the last one hundred years or so, it has become apparent that the world of 'matter' has also developed in complexity over vast periods of time, and the term 'evolution' is also used to characterise this development. It now appears clear that there is no sudden distinction or dividing line between evolutionary processes in the inanimate sphere and those in the sphere of life, nevertheless the terms in which these two spheres are generally understood are so different that it can be difficult to conceive of the connections there may be between them.

At the time Darwin began his research it was already

becoming clear to at least some observers, that rather than being fixed for all time, species of creatures were subject to gradual transformation of various kinds, over generations. There were several different theories concerning this, but Darwin and Wallace's detailed inquiries revealed important aspects of how this transformation comes about, and the overall principles they outlined became established as the basis for evolution theory as we know it today. The phenomenon was seen for many decades to be something that related to living, but not inanimate, nature and to a considerable extent that perception continues. However, discoveries in the fields of sub-atomic physics and cosmology have revealed that ongoing and continuous developmental processes are occurring at sub-atomic and cosmic scales also. These processes are also referred to in terms of evolution, though this is generally understood in terms very different from those applied to living processes. The use of the term 'evolution' in this context is generally referred to as 'cosmic evolution', its central principle being known as 'nucleosynthesis', which I'll outline presently in a rudimentary way. However, although these two uses of the term 'evolution' may appear very different at first glance, it has become clear that no fundamental dividing line appears to exist between them at the level of physics or chemistry. At those levels, the only clear difference appears to be the degree of coherent organisation, or complexity, involved. My interest is in seeking common principles of evolution in terms that make sense across our view of developments so far, from the level of primordial universal origins to that of conscious human experience. This broader, inclusive concept of evolution is what is referred to here as 'universal evolution'.

In these discussions I present the view that *all* kinds of

inquiry and study are most mutually comprehensible when viewed within the perspective of evolution, as this provides the ultimate backdrop to all specialisms. To this end I outline and explore to a degree, some simple overall principles of evolution as best I can, and later, human consciousness and culture as a domain within the broader scope of evolution, which is particularly relevant at this time. The principal observations and perceptions I offer here are not new and are intended to be in pursuit of useful understanding rather than purely theorising for its own sake...

The Classical Science View of Evolution

For those unfamiliar with the great shifts that have taken place in scientific views over recent history, the next few paragraphs may be worth reading a couple of times...

In the mid-19th century, when Darwin was making his observations, science had as yet no clear knowledge of the atom other than as a concept of a minimal unit of matter, none at all of internal nuclear structure. Also, Einstein had not yet appeared and expounded his ground-breaking theories and science was still firmly based on terms which we now refer to as 'classical' physics theory to distinguish it from the radically different ideas later ushered in by the age of relativity he delivered, followed just a few years later by those of quantum physics which was initiated by Niels Bohr and others (following up on earlier work by Max Planck). The 'classical' view was largely laid out by Newton in the eighteenth century. Essentially, this holds that every effect has a cause, and logical inquiry can, in principle, reveal the causes behind all events. This was a tremendously effective approach to inquiry which had developed its methodology, and its philosophical basis, largely over the preceding two centuries*1 and it still bears great influence in today's thinking because it relates closely to normal everyday experience.

This approach was ultimately to provide the tools, theory and methodology that came to underpin the industrial revolution which has directly and indirectly transformed the nature of civilisation as we know it, opening the way to all forms of mechanisation, mass-production, heavy engineering, motorised travel, electronics, radio communications, air travel and exploration of space. Science in this form has been so effective however that many of its concepts have become established in the Western mind-set as reality per se, rather than as *descriptions* of reality. Some of these conceptual creations were later revealed, implicitly, by Einstein to be purely relative notions, but unfortunately his theories are not easy to grasp, even for most scientists let alone the general public. In effect, Einstein demonstrated that many ideas which we regard as being 'true' representations of reality - including concepts like time and space - are nothing of the kind, but are mental constructs which we have come to identify as reality. Einstein is by no means the only person who has done this in one way or another. Many have, before and since. In his case though, great attention has been paid to what he said because he spoke in the practical and contemporary language of science, supported by mathematics, to a technically receptive audience rather than the language of philosophy or doctrine to a much wider and generally less discriminating audience, which has been more usually the case. However, because of this context the focus has been primarily on the 'objective' significance of his observations, with little attention paid to 'subjective' implications concerning the nature of abstract thought. Further investigation into the nature of matter and energy at the atomic level has revealed features of behaviour that are largely incompatible with both the classical view and relativity, giving rise to quantum theory, which has its own entirely different and unique methods of inquiry and descriptions that now underpin much of modern technology, but conflict in significant ways with the classical understanding of causality. In practice though, the classical view is still predominant because for most practical purposes it is still more directly accessible and better reflects our intuitive sense of logic.

When Darwin was making his observations and developing his ideas, the classical view was the only available scientific perspective, and his theory was adopted entirely within the terms of that perspective. This is normal of course, this is how we develop knowledge, but as we do so its base and scope constantly expand. Thus does our collective knowledge and understanding grow and develop. The only certainty in knowledge is that we will come to know more, and current knowledge will become outdated... This doesn't mean that current knowledge is untrue, but it does mean that it is *relative*, because we are always bound to interpret observation in available terms. There are no absolutes in measurement or description, and as we find out more, our understanding of evolution evolves also. The terms of understanding that prevailed when Darwin presented his theories have moved on significantly, and we are obliged to re-examine those terms to bring our understanding of evolution up to date.

Let me introduce the reader to some areas of inquiry that are particularly relevant to obtaining a wider conception of 'evolution' than may be more usually encountered. I don't intend to go into these areas in depth, as my purpose here is just to illustrate their common basis and continuity without getting too complicated. Accordingly, the references I make to scientific background information here are only intended to provide a broad outline of relevant features. More detailed information on aspects the reader may want to follow up on is easily available these days of course, online.

Cosmic Evolution

Many apparently complex ideas that have come into common use have origins in everyday observations carefully explored. The notion of the universe beginning in a 'big bang' originated in just such a way. The background to this theory involves several steps, starting with the rainbow...

The following is a very brief outline.

In the 16th and 17th centuries several scientists - best known perhaps, Isaac Newton - explored the structure of sunlight using prisms which revealed that it is composed of several distinct colours, which Newton called a spectrum. This was by no means a new observation, but Newton greatly contributed to analysing and elucidating its nature. Later exploration, in the early to mid-19th century particularly, began to reveal finer details in the structure of the spectrum, concerning interference patterns in the light emitted by different substances in combustion, which offered a method of identifying elements present in its production. This exploration of the behaviour of light underlay the emergence of spectroscopy as an important analytical technique. As understanding of the vibrational quality of light grew, this had considerable implications for different spheres of the sciences. On one hand it provided a powerful new tool for chemical analysis and insight into the electromagnetic spectrum generally, at the same time it became useful in the field of astronomy, in exploring the chemistry of stars. This made it an extraordinarily useful technique, considering the immense distances involved.

Concerning these spectral patterns, in 1842 Austrian physicist Christian Doppler described a phenomenon whereby the frequency of a waveform changes in relation to the position of an observer. Again, this is commonly experienced on an everyday basis, as for example, when a fast-moving vehicle approaches and passes us, the sound appears higher on its approach, lower after it has passed. Light does essentially the same thing, and this is represented in the appearance of spectral patterns. If an object being observed is approaching, its overall spectral pattern is raised in frequency, known as 'blue-shift'. If it is receding from the observer the frequency is reduced, called 'red-shift'. Accordingly, this phenomenon has come to be known as the Doppler effect, or shift.

Early in the 20th century, with the development of high-powered telescopes and spectroscope technology, it was becoming apparent that the universe is expanding. This was particularly evidenced through the observation of the Doppler effect, principally redshift, in the movement of distant galaxies. It could be seen that galaxies in every direction we observe are moving away from ours, not only that, but they are constantly moving apart at a rate related to the distance between them - the further apart, the higher the velocity of their separation from us. Effectively, the universe is seen to be expanding. This expansion was identified, and its rate calculated by Edwin Hubble in 1929, known as Hubble's Constant (though there are other important factors involved in addition to the Doppler effect, not least, relativity, but the detailed observation of the behaviour of the light spectrum remains a key factor). Prior to this it had been generally assumed that the universe existed in a more or less 'steady state' and had done as far as we could know, for ever. Although it took some time to digest its implications, the discovery that the universe is expanding introduced the recognition that if the galaxies are receding from each other, clearly in the past they must have been closer together. There were competing theories regarding this, but the overall winner eventually proved to be what has become known as the 'big bang' theory.

The 'big bang' theory draws on cosmic observation of this kind to conclude that the universe as we know it emerged spontaneously around 13.8 billion years ago, from a 'point' completely without dimensions, as we understand that term, which is referred to as a 'singularity' (one shouldn't try too hard to conceive of what a singularity is because ultimately the term simply means a state beyond all definition in terms of any of the parameters we normally use for mental conception or measurement, including time and space, and matter and energy as distinct entities - in other words, beyond conception - and in that sense it simply means 'the beginning of time and 'space', which are conceptual parameters). This took place in an event of unimaginable power and intensity which we compare to an explosion, but one in which space itself expands at an enormous velocity that continues to this day, which we can see evidenced by the 'red-shift'. As this proceeded, the original concentration of extreme energy dissipated, allowing the cosmic temperature to drop, eventually allowing the coalescence of primordial particles, creating matter that we are familiar with in the form of atoms. This is the basic idea (greatly simplified) that now largely underpins our modern view of cosmic evolution, in contrast to the 'steady state' theory, which effectively represented the classical view of universal processes - essentially as forever constant - incorporated into the developing cosmic theories.

As far as scientific theory is concerned, whatever may have 'preceded' this state is beyond our powers of conception and our skills of empirical inquiry, for which reason attempts at *all* theorizing into that aspect of existence must be viewed entirely as conjecture and speculation. Although scientific inquiry depends on the exercise of imagination and deduction, it also requires these to be constrained by evidence, or at least, recognised as such - as unsupported products of imagination. For that reason, science concentrates on the 'observable universe'. This begins for us in the earliest instants of this emergence of 'primordial substance'. At this point, there were no atoms as we know them yet, nor even protons. Atoms depend first of all on the presence of protons, and at this stage even protons had yet to form.

To explain a little...

In earlier days of science (*and initially conceived of by the Greek philosopher, Democritus*) atoms were proposed as being the smallest possible constituents of matter, but in more recent times it has become clear that they are in fact composed of several more elementary particles - protons, neutrons, and electrons. More recently still it has come to be understood that for the most part, even these particles are composed of a whole range of even more fundamental constituents. Whole families of such constituent components have been explored and described, and this exploration continues...

The following outline description is based on my (very limited) understanding of current theories concerning cosmic and sub-atomic processes. These are at opposite ends of the scale of observation but intimately related. This outline is very approximate.

In the first moments of existence of our universe then, as conceived of in the 'big bang' theory, atoms which compose the world as we know it, did not yet exist, only elementary sub-atomic particles which were dispersing at high velocity (approaching, or possibly greater than, the speed of light according to some interpretations) in all directions, viz the term 'big bang'. The temperatures prevailing at this time are reckoned to have been in the billions of degrees Celsius, and elementary components of matter, such as quarks, which combine to form the central ingredients of the atomic nucleus would not yet have been able to do so under those conditions. But as the universe expanded and its energy became more widely distributed this temperature would have dropped, gradually allowing conditions for those most primordial particles to coalesce to form the first atomic nucleic particles - protons and neutrons. The first 'complete atoms' came into existence as these particles engaged with electrons (which belong to another branch of the family of fundamental subatomic particles) as cooling continued. The simplest and first atom to appear, consisting of one proton and one electron, is the element we call hydrogen. This element is widely distributed throughout all observable cosmic processes and events and is central to the formation of stars and galaxies. Around this time helium, consisting of 2 protons 2 neutrons and 2 electrons also began to appear. At this stage only a small number of further elements were able to form, and in smaller quantities still.

Later, gravity which effectively is a force of mutual attraction associated with the mass present in most constituent atomic particles - possibly preceded by electrostatic aggregation, another principle of attraction - drew vast quantities of these initial elements together to form the first stars. These stars, initially composed mainly of hydrogen plus a much smaller amount of the other elements touched on above, formed the engines of production of a further generation of more complex elements through the fusion of nuclear constituents - protons and neutrons. Later, depending on their size, these stars would collapse and dissipate in a variety of different ways, or explode as nova or supernova, as they reached the end of their available fuel for fusion. The new elements dispersed in this way then became available, again under the influence of gravity, to contribute to the seeding of a new generation of stars and the planets around them. These second generation stars were then responsible for the creation of further elements. This process appears to have been repeated through many such generations, with the lifespan of succeeding generations increasing as the complexity of their participating elements increased. Our own sun is a relatively recent arrival in this sequence of events, with its initial composition represented also throughout the solar system at large, including of course, planet earth - all originating from the debris created by the destruction of previous generation stars. Thus, have ever more complex elements been created. This is a brief and simplified outline of the process of cosmic evolution of the elements, known as *nucleosynthesis*, from which our physical existence is constructed. The material from which the sun, the entire solar system including planet Earth, and consequently us, is composed, appears therefore to have been formed in this way, around 4.5 - 5 billion years ago as the cumulative result of the work of stars of several, probably many, previous generations.*2

Mineralogical Evolution

The story of cosmic evolution explains to a great degree the existence, origin and evolutionary relationship of the many elements that exist in the universe, but further than that, we now know much more also, about the wide variety and forms of minerals which compose the structure of our planet, and the ways in which these have developed through many and ongoing, degrees of change. Minerals are composed of many different combinations of elements. These combinations of elements have cosmic origins in the 'end of life' and dispersion of the material from previous generations stars, and to a considerable extent they are also influenced by the ongoing very dynamic relationship the sun has with everything within range of its heat and other forms of radiation. The variety of these combinations of influence has continued, prior to and throughout, the formation of the Earth. This formation and transformation of minerals involves a wide range of chemistry, which includes the production of complex organic molecules and compounds, many of which are also present in the chemistry of life. We now know that these processes therefore began well before the stage at which the Earth even existed as a planet and continues by the constant re-distribution and recombination of materials through volcanic and tectonic processes. More recently in geological timescales, living processes have augmented these processes by introducing further dimensions to those changes, such as for example, through the liberation of atmospheric oxygen, which feeds back into the molecular chemistry of the planet. This has added to the variety of minerals and compounds which exist, and which participate in the convectional recycling of our planetary material, producing further complexity in the domain of mineral content throughout the planet. The relationship between mineral structure and living chemistry is clearly very intimate and of great interest in current exploration but is far from being fully understood. Awareness of the contribution that the complexity of mineralogical development makes to evolutionary processes is relatively recent and only now growing and should be considered a sphere of considerable relevance in its own right.*3

Organic Evolution

Of course, this is the aspect of evolution that we are most familiar with and probably needs little introduction other than to remind the reader of the central and best-known principle, described by Darwin - that of 'natural selection', which will be considered further in the next chapters. This is a process whereby nature generates wide variation, which through inheritance and ecological 'competence' establishes the predominance of 'best fitted' individuals and consequently species, in any sets of circumstance. (Although the expression 'survival of the fittest' is commonly associated with natural selection, and Darwin approved of it, it was originally coined by Herbert Spencer, an economist, not Darwin himself.)

Another important point that arises from inquiry into organic evolution, particularly from genetic studies and observation of cell structure, is that on account of certain essential features common to all life forms that we know of on Earth, these must derive from an initial common ancestor, which is often referred to as LUCA (for 'last universal common ancestor'). It is clear however that this common ancestor couldn't have just sprung into existence, but emerged as a result of prior evolutionary processes, during an era of transition from inanimate molecular chemistry to that of life as we know it. The nature of this transition is a matter of great interest and inquiry.

This theory also represents the basis for the understanding of ecology, which expresses the intimate balance of relationship between all living things on our planet. I'll be discussing this further in a later chapter. Although we now tend to think of organic evolution in terms of genetics, at the time of Darwin's observations of natural selection, nothing was known about genes per se and some of his observations relate to principles that can be seen to be expressed more widely in a simpler form in inanimate nature also, although this couldn't be recognised at that time. This topic is integral to many aspects of the ideas I will present.

These three areas of evolutionary process are the principal ones recognised as such, in one way or another, by science currently. I will later add to them 'conscious cognition' as a further area, with its own characteristics which I'll also discuss, and 'culture' as the currently dominant sphere of evolutionary development on earth...

But before all of that, I'll discuss further, the concept of evolution as a universal phenomenon.

Chapter Notes

*1 since the mid-16th century, when Copernicus got the ball rolling, as-it-were, with the publication of his book, 'On the Revolutions of the Celestial Spheres', which placed the sun at the centre of the solar system. This disrupted the ancient view of the earth being at the centre of the cosmos and effectively set the stage for the emergence of the new, now classical, physics *which was to a large degree established by Newton. At the centre of this view was the principle of causality; cause and effect - every event has a cause. Subsequently, relativity theory made clear that the position of the observer is central to the way in which cause and effect are related and interpreted, and quantum theory has made it clear that at fundamental levels of nature, causality as we know it at larger scales doesn't exist, at least in terms that we are familiar with.*

[2] at the earlier stages of stellar process, when the composition was simpler, and depending on their mass, the life span of stars would have been much shorter, sometimes perhaps as little as 100 million years or less, becoming longer with each generation, and, depending on size, which is an important factor in the 'lifestyle' and duration of stars. That means there has been time for many generations preceding the arrival of our sun.

[3] my introduction to the importance of minerals in the scheme of things came through reading some of the work of Robert Hazen, professor of Earth Sciences at the Carnegie Institute of Washington.

Chapter 2

Universal Evolution

Outline

As I touched on in the introduction and the first chapter, certain underlying assumptions remain within modern day thinking that have their roots in ideas, or views of the world that have now become effectively obsolete because of more recent inquiry and discoveries made, particularly over the last couple of centuries. Several of these assumptions still bear a considerable influence on our ways of understanding the workings of nature and evolution however. It will be necessary and interesting to be aware of and explore some of these at points during this inquiry, but I'll focus first on presenting here an outline of the central ideas I want to offer.

In this section I present the view that all levels of phenomena, animate and inanimate, tend to spontaneously and coherently 'self-organise' according to certain common principles that are to a degree identifiable. From this perspective, domains of natural process that we tend to think of as isolated areas of inquiry are all subject to those same principles, whether they be at the level of the sub-atomic, molecular, mineral, organic, conscious cognition, cultural or anything in-between - and the same ones that underlie self-organisation are also involved in the processes of disintegration.

Viewed in terms of self-organisation, the attention is drawn to the way all individual phenomena exist within an environment which, in conjunction with the qualities of the participants, determines their behaviour. All events are engaged in some form of dynamic system, in which they are governed from 'below' by their more ele-

*mentary constituent factors and 'above' by more complex environ-
mental influences, or we might say, from 'within' and 'without',
where 'within' refers to internal constituents and 'without' refers to
environmental factors. This leads to the observation, among others,
that there are no hard boundaries between apparently very different
levels of complexity of universal processes other than those introduced, or perceived, by the observer; rather, there is an unbroken,
fluid, chain of relationship and interaction.*

Predating the 'classical' scientific view which I mentioned
in Chapter 1, inquiry into the workings of nature was known as
'natural philosophy'. As scientific inquiry developed and diversi-
fied, this term became identified with the 'cause-and-effect' ethos
of the classical approach and the field of physics in particular.
Later, it fell out of use as an outdated concept, leaving that of
'physics' to represent the primary approach of the sciences
generally. However, I would say that the concept 'natural philosophy' relates more to the focus of attention on first principles of
natural behaviour rather than to the detailed analysis and complex
theories that came later. Some of the terms I use in this chapter
don't conform directly to current conventions of science, but may
perhaps relate more to this earlier view, drawing on relatively
simple observation for inspiration. It may be worth bearing in
mind that all inquiry begins with exploring the obvious...

Self-organisation

Self-organisation is now a well-recognised phenomenon in
the world of science, where it is understood that spontaneous organisation takes place in many contexts and at all levels of nature.
It is a basic feature of all molecular organisation, and prior even

to that, of the interactions of sub-atomic particles. It's a funda-mental feature of all forms of collective behaviour - including the social behaviour of organisms. It isn't too much to say that without this principle of nature, nothing that we know of, from atoms to ourselves, could exist. Spontaneous self-organisation could be said to begin at the very point of energy transforming into matter, and is present throughout all forms of organisation, including conscious human activity.

Ever since the beginning of the industrial revolution, the attention of scientific inquiry has tended to place emphasis on the tendency of natural processes to proceed in the direction of disorder, scarcely recognising the phenomenon of self-organisa-tion. This orientation was established at a time when the priority of scientific inquiry had become one of controlling energy from an engineering perspective, finding its most complete expression then in thermodynamic theory. The recognition that all forms of order in nature arise spontaneously through inherent principles came more recently, particularly over the last couple of centuries (in Western culture that is, certain Eastern cultures have long rec-ognised this). From a technological perspective, self-organisation is largely taken for granted and invisible, but in forming any un-derstanding of the workings of nature, recognising and exploring this principle seems essential. Its importance is both philosophi-cal and practical because the same essential factors of self-organ-isation are present in every sphere of nature.

The principal difference between self-organisation as it is expressed at the inanimate levels of nature and that of living organisms, is the ability of living organisms, however simple, to reproduce molecular constructs in a much more systematic way that leads to the development of much greater complexity than can be obtained by atomic/molecular principles alone. Nevertheless, it seems important to recognise that self-organisation is as much a feature of inanimate nature as it is of life. Significantly, even the world of apparently basic matter that surrounds us only exists by virtue of the emergence of its component parts through the self-organisation of the most elementary primordial constituents, which of course continue to exist and underpin everything else.

Principles of Self-organisation

In a later chapter I'll discuss the position of the observer in relation to the view perceived, so I won't go into that matter too much for now, but it does need to be recognised early on that some important assumptions in our way of interpreting the world of events, which were established centuries ago when our knowledge was much more limited, still influence our understanding of nature in important ways. Some of these assumptions need attention. My intention here is to shift the view of universal processes away from one in which the perception of 'randomness' tends to dominate, to one that focuses more on the universality of coherent relationships as a principle throughout nature. The term 'self-organisation' refers to the phenomenon whereby nominally independent entities and forces engage and interact spontaneously according to underlying natural principles to produce all the forms of order, organisation, and integration that we can observe at all scales of natural events, without the requirement of any additional form of control or manipulation. This is by no means a controversial view, the whole of modern physics theory rests on the assumption of coherent relations of this kind; sub-atomic particles combine coherently to form atoms; atoms combine coherently to form molecules; molecules combine coherently, creating the vast range of minerals that comprise our planet, and, within the organic environment, provide the building materials of all organisms. It becomes apparent then that phenomena such as life, consciousness, and intelligence, as we perceive them within and around us, have evolved, as-it-were, microscopically and in a coherent and integrated way, upon the broad basis of more elementary domains of evolutionary process, rather than as secondary, chance consequences of entirely random processes.

Coherence and Randomness

I want to divert briefly to discuss the idea of randomness because it is an interesting concept that insinuates into many different situations and has a prominent role within popular and scientific thought. As with many concepts when considered closely, this term can have several subtly different meanings depending on its context of use. This can range from the informal context of everyday experience to the formal context of mathematics where it relates to factors of probability. Very broadly, the concept 'randomness' can be seen to derive historically from our everyday perception of two quite different features of behaviour that are frequently confused or conflated to some degree, which concern *predictability* and *purpose*. When we don't know the background to events and/or they don't appear to have been intentionally initiated, we are likely to perceive them as occurring 'randomly'. For example, a rock falling onto a road from a cliff face is generally perceived as random, unless we should see someone causing it to do so. If we have some knowledge or experience of the causal background, or if we consider it to be the result of intentional action, we would generally consider such an event to be non-random. The point here is that in either of these cases, the perception of 'randomness' reflects the status of the observer rather than that of the event itself. This raises the further question of whether 'randomness' truly exists as an objective feature of reality, beyond the sphere of subjective perception. This has been a matter of debate for many decades if not centuries and continues; I don't intend to explore that topic for now, my main interest lies in the *perception* of randomness.

My observation here relates to predictability rather than intentionality. In contrast to randomness, regularity and consistency are qualities we also experience. We could refer to these qualities as *coherence*. In terms of natural behaviour, both randomness and coherence are innate features of our everyday experience and observation, and generally co-exist in an amicable way. For example, the Earth revolves around the Sun, creating the seasons with generally predictable, seasonal, qualities but day to day, even hour to hour, conditions are much less predictable, or random. We experience that unpredictability but recognise its inclusiveness within the context of a greater coherent pattern. A variety of studies of modern science have provided us with knowledge which has made it clear that principles of nature direct events in coherent ways even when these events appear random in terms of our everyday experience. From this point of view, while the perception of randomness is undoubtedly valid at the level of gross experience of events, this doesn't conflict with the inclusive coherence of natural laws that underlie and govern those events. Even disintegration follows the coherent instruction of natural law. From this perspective, as a principle, coherence *precedes* randomness in significance. This relationship between randomness and coherence is at the heart of natural selection, which I'll discuss.

I would say that an emphasis on the perception of 'randomness' as an objectively existing quality of nature has been greatly enhanced by the influence of thermodynamic theory on Western thought, in which entropy, or the tendency towards disorder has become a dominant feature of our view of universal processes. In my view, this is a result of the focus of attention on the investigation and control of energy which has helped to drive the industrial revolution over the last three centuries. This isn't intended to question the validity of the concept 'entropy' but rather, to look beyond it to the coherent qualities of nature that precede and underlie it.

Ingredients, Environment and Complexity

We now know that the universe didn't begin with complex structures such as life, galaxies, stars or even atoms - at all stages, an evident feature to be observed of events throughout the universe is that of growing complexity based on a combination of the internal structure of participating entities and external environmental factors and forces present in the areas under observation. At every observable level of evolutionary process, it can be plainly seen that far from being essentially random, or arbitrary and fundamentally disorganised, the behaviour of all the composite elements of structures observed is a result of the coherent interaction of those component parts with each other. Another important implication of this observation is that at every level of organised existence, all component entities whether they be conceived of in terms of atoms, molecules, organisms or 'minds', each is the result of prior complex processes and all share a common evolutionary status. What I mean by this is that none is more or less 'real' than others.

All interactions involve two main aspects; the first of these concerns the nature of the internal organisation or 'structure' of individual participants involved, the other is provided by the forces and circumstances present in the environment, including all other participants. This relationship between 'inner principles and the 'outer' environment is central to the direction that self-organisation takes. For instance, sub-atomic particles such as quarks combine in consistent ways to produce higher level sub-atomic particles, i.e., protons and neutrons, which along with electrons are the principal components of atoms. These are then subject to the forces and 'fields' which govern the more macroscopic molecular domain, particularly electro-magnetic energy and, again, gravity. This overall relationship is one, which has been progressing since the earliest stage of the known universe's existence, through degrees of ever - increasing complexity.

A simple example that illustrates this principle of the relationship between the established internal composition of a component entity and its environment is a well-known experiment which many readers will have encountered in a school science lesson, whereby iron filings are sprinkled onto a sheet of thin card covering a bar magnet. In this experiment the filings arrange themselves into a pattern that corresponds to and creates a display of, the magnetic field produced by the magnet. This example serves several useful purposes.

The first of these is this; while the relationship of the filings to each other in the absence of the magnetic field may reasonably be characterised as 'random' (from the observer's perspective at least), in the presence of the magnetic field the arrangement that they adopt tells us something about both their own internal structure and the nature of an influential 'force principle' present in their environment. Each individual filing, minute though it may be, is the result of a prior complex history of cosmic evolutionary processes. As an atomic element, iron, it has properties and internal structure which are particular to that element. However, in terms of the overall situation under observation we may regard it as a participant or 'component'. Meanwhile, in addition to the individual 'components' represented by the filings, there is the 'environment' which bears a significant influence on their behaviour, namely the magnetic field of the bar magnet (there are many more environmental influences present, not least, gravity, but for the sake of this example the predominant one is the influence of the magnetic field). It can also be said that under the influence of that field the filings collectively become participants in the 'environment'. On the one hand they respond according to their own internal structure, at the same time they interact collectively as mutual mediators of the environmental influence, the magnetic field. For example, one effect of the magnetic field on the filings is that they become individually magnetised which, as they are then all polarised the same way makes them keep their distance from each other in certain ways, rather than forming an amorphous mass.

While an observer perceiving the arrangement of the filings may interpret this in 'static' terms, it would be more realistic to see it as essentially dynamic. Simply, the magnetic field can be considered to exert a forceful environmental influence and this force governs the individual and collective behaviour of the filings.

Now, taking this brief outline and applying it to other

kinds of situations and levels of observation, certain comparisons can be made. At the sub-atomic level elementary 'particles' can be said to have their own distinct 'internal' characteristics, which respond and interact dynamically under the influence of the forces and fields of their domain environment, such as the weak and strong nuclear forces. This creates higher orders of structural complexity, i.e., protons and neutrons, the nuclear components of atoms, which then combine with corresponding numbers of electrons to create what we regard as more complete atoms (again, in terms of cosmic evolution, this higher order was initially represented mainly by hydrogen which is a single proton with a single electron). These individual entities of higher order structure then interact under the dynamic influence of their environment, creating the domain of molecular organisation.

This same pattern of relationship between the contributing 'components' of organised systems and their environment can be recognised at any level of evolutionary process one cares to observe, from the inanimate, through the animate, from that of sub-atomic particles to that of conscious interaction and human society/culture, and is at the heart of spontaneous self-organisation in all spheres.

These various levels may appear to be entirely distinct, but close inspection reveals an unbroken chain of accumulating degrees of complexity, each stage of which supports and is integral to the next. At each stage the behaviour of 'components' relative to their 'environment' is perceived and interpreted by us in terms of the apparent dominant features of that stage. We refer to the collective behaviour of sub-atomic particles in terms of strong and weak nuclear forces; of atoms, elements, molecules in terms of electromagnetic forces; of stars and galaxies in terms of all of these, but with particular attention to gravity because of its central contribution to the processes of nucleosynthesis. Once we reach the consideration of organic processes, we observe the interactions of 'components' in the form of individual living entities, which for example may consist of bacteria, insects, trees, human beings or any other organism. Their 'environments' can be viewed in terms of complex ecology which, as well as the organic circumstances, includes all other contributing factors, such as climate and geography, and in the case of higher evolved species, the environment of conscious cognition. As observers we may consider these areas as distinct evolutionary domains, but it should be remembered that all are entirely and necessarily cohesively integrated, and I'll make some observations about the nature of that integration presently.

While the specific terms that may be appropriate in observing and interpreting events can clearly be very different depending on the levels or areas being observed, certain general features of the relationship between the 'components' and 'environment' being observed remain common to them and can be recognised. Three main factors of this relationship can be characterised as; *energy*, as a principle of action or dynamism; *equilibrium*; and *freedom of movement*, which I've touched on. These are already well-known characteristics of course, but usually conceived of and described in the terms of classical thermodynamics, where the focus of attention is on the control of energy transactions. Viewed in terms of self-organisation however they can be seen as having a much broader, more holistic, meaning that reflects the complete and intimate nature of the relationship between component entities and their environment. Again, a quality that can be said to include, or characterise the relationship between all these others is *coherence*.

In modern science for instance, the use of the term 'energy' generally relates particularly to its adoption as a principle in thermodynamic theory. In that context it represents a principle of force that can be harnessed to do 'work'. In other words, rather than being an actual independent principle of nature, it is really a concept we have adopted to convenience and grasp the relationship between different aspects of the behaviour of 'matter'. As a comparison, we use the term 'field' in a similar way to help visualise the way in which electromagnetism and electricity propagate. These terms are very useful but at root they are essentially metaphors of description in terms we can relate to. 'Energy', in this sense, isn't something that exists on its own. We conceive of it in terms of transactions we can perceive in the domain of physical processes. The same applies to the concept 'matter' of course, thus we come to talk and conceive of these transactions in terms of 'energy' and 'matter' as if these are two separate things. Indeed, we did think of these as different things until it became clear early in the 20th century that they are really two interchangeable aspects of reality (again, courtesy of Albert Einstein). However, this use persists on account of the convenience it offers in interpreting physical transactions.

For my purposes I draw on its meaning as a 'principle of force' applied more broadly, beyond the scope of purely physical expressions of self-organisation, into other regions, such as animate motivation and domains of conscious experience. What I mean by this should hopefully become clearer as the discussion proceeds.

Similarly, equilibrium, or stability, shouldn't be thought of in static terms. In all realms of evolution, the development of complexity is entirely dynamic, regardless of its pace. Stasis is an illusion of perception. This is no doubt necessary for our practical everyday existence, but the behaviour of every atom or sub-atomic particle is governed by its environment to some degree, locally and universally, and this is always in dynamic transition of some form. For this reason, a rock, for example, may appear to be entirely static and, left to its own devices, permanent, but in terms of cosmic processes and timescale, however conceived, its form is transient. Even a proton may have a life span, though that could be many billions of years. Equilibrium as a concept relates more meaningfully to the balance between the elements of a dynamic system than as representing a truly 'static' quality; thus, an atom represents a stable state of relationships between its sub-atomic component parts, a molecule is a balanced engagement of component atoms and so on. It is implicit therefore that equilibrium is a relative term, and its representation within any form of complex organisation always reflects the nature of the relationship of participants with their environment. In the context of spontaneous complex development - i.e., evolution - stability, or equilibrium, represents the ongoing thread of continuity that supports and is central to that development. This thread is fundamental to evolutionary organisation in all domains, the source of strength in general, but fragile at any given point, and never static, always drawing on and subject to influences of the environment.

The third principle here is 'freedom of movement'. From the discussion of energy and stability, 'freedom of movement' may seem irrelevant or at least, rather obvious. The reason it has to be considered is that, at an everyday level, and in a significant way our interpretation of this quality tends to be to characterise it as 'randomness'. This concept carries strong implications of 'arbitrariness', which conflicts with the coherence, and regularity of nature's constant tendency towards equilibrium in one form or another. As I've been outlining, at every level of complex organisation in nature there is a relationship that can be observed of the individual 'entity' to its environment, and 'freedom of movement' is essential to that relationship. Without that 'freedom' interaction could not take place. However, the idea of 'freedom' here doesn't imply 'freedom to do what it likes', rather, it suggests freedom to move or respond to environmental influences. A simple everyday example of the effect of freedom of movement that also acts as a general analogy to the relationship between these three principles is that of water, which we commonly know of in three different forms: ice, liquid, and vapour. In the form of ice its structure is very stable but relatively unresponsive or inaccessible to environmental influences other than temperature and gravity; as vapour it is unstable and incapable of forming, or contributing directly to, the formation of stable complex structures, but in its liquid state it can dissolve, contain, and interact with a wide range of elements and compounds, thereby contributing to the formation of a wide variety of molecular constructs.*[1] We can see these three states as exemplifying degrees of 'freedom of movement', which are expressed in a wider sense throughout nature, between the extremes of freedom and restriction at either end of the spectrum of universal temperatures. This represents the general principle that between 'too hot' and 'too cold', or too rigid and too flexible, we find degrees of 'freedom of movement' that are supportive of evolving complexity. This condition is an essential requirement in all domains.

In modern times, it has become common to speak of the evolution of complexity in all its forms, physical and abstract, in terms of 'information'. It's not a term I'm fond of personally, but from that perspective it's clear that the same basic rules apply throughout all the range of forms that complexity takes. Although the principles of energy, equilibrium, and freedom of movement (as randomness) have originated and been derived from the observation of, and inquiry into, inanimate physical behaviour, they aren't confined to such considerations, but can also be seen to be represented in living processes and less tangible domains of organisation, including those related to conscious experience.

Summary

Albeit, perhaps a little unconventional, I don't think that there is anything controversial about the observations I have been making, but having 'loosened the ground' a little, I can now outline the premise of evolution as a continuous unbroken and self-sustaining process, from the 'beginning of the universe' to 'now'. Based on the foregoing there are several main summary observations I make to support this view.

The first and most important is that nothing we know of exists in complete isolation from its surroundings, and the nature of this relationship, and all relationships, is dynamic not static. Stasis is an illusion of analytic observation. Another is that the 'classical' interpretation of the development of complexity rests to a great degree on the notion of 'randomness' as a primary aspect of physical interaction, e.g., sub-atomic particles bump into each other purely by chance. But as I have discussed and science has demonstrated, all observed relationships in all domains of nature appear to express coherent underlying integrity and conform to consistent rules of behaviour (though we don't have full knowledge of those rules). From this perspective 'random' doesn't mean arbitrary. It doesn't take too much adjustment of view to consider that at each level of natural activity, the 'components' of that activity are bound to interact with their environment in accordance with their internal structure and the qualities and influences of that environment. This applies whether those 'components' are sub-atomic particles within atoms, at one end of the evolutionary spectrum, or human beings within society at the other. It appears clear to me that this relationship is far from random (though at this point I am only interested in this coherence, not in speculation about matters such as determinism or intentional creation). Without such a coherent basis, even atoms could not have come into existence. The view that emerges here is primarily one of coherent, dynamic, self-organisation in which cumulative complexity is the basis of all evolutionary form, i.e., everything that exists, from sub-atomic 'particles' to the domain of conscious cognition.

Chapter Note

[1] it's worth noting that although our most common perception of water is as a liquid, it can only take that form, normally, between the temperature extremes of 0 and 100 degrees Celsius approximately. In terms of universal temperature scales, this is a miniscule range. Without its presence in this form life as we know it - or possibly at all - couldn't exist. The importance of water to the existence of life both as an ingredient, and as a medium of influence can't be overestimated.

Chapter 3

Domains and Hierarchy

Outline

My use of the term 'domains' is as an informal way to distinguish broadly between different levels and regions of evolutionary complexity on the basis I have been discussing, and shouldn't be confused with its established use in biological classification. This use isn't intended to be very technical, but simply to convey a sense of the integrated and progressive relationship between apparently different levels of evolutionary phenomena, while recognising that each can be viewed in distinct ways. The term 'hierarchy' is used to indicate that complex organisation is governed from 'top-down' rather than 'bottom-up'.

If we think of evolution in terms of self-organisation, in which the progressive development of complexity is apparent, certain consistent aspects of this development can be seen to be represented across different areas of observation. This I've touched on already in examples such as the relationship between subatomic particles and atoms and that between atoms and molecules/compounds. We may regard these different areas of observation as particular regions of complexity, or 'domains', each of which may be viewed in terms of its own 'rules of behaviour'. As far as evolution is concerned however, these domains are entirely integrated, continuous and interdependent, each being supported from below by less complex, prior, domains. Again, this is exemplified in the way the domain of molecular structure is supported by that of atomic structure, and the atomic by the sub-atomic.

This ascending relationship is maintained throughout all realms of evolutionary organisation. It is for the observer to adopt a viewpoint that allows an overall perspective of that continuity and interdependence. Sometimes, as in the examples I've mentioned, this relationship may be reasonably clear and comprehensible, sometimes much less so, as in that between the inanimate domains and the organic. Through the sciences of physics and chemistry much has been achieved in understanding some of the underlying 'mechanics' involved, but the range and terms of that understanding don't suffice to provide complete explanation beyond that, into the sphere of living processes.

However, it seems to me that given the information we now have about the various domains, inanimate and animate, certain overall patterns of relationship can be recognised on broader terms than they could previously, and I've begun to discuss this in the last section, around self-organisation. These patterns largely reflect the essentially dynamic status of all natural processes and events, rather than the detail of 'components', though those remain entirely relevant.

Domain Hierarchy

While individual 'components' such as sub-atomic particles, or atoms, have their own particular characteristics, and like dancers in a dance, interact with their environment in consistent ways that reflect their innate 'physiology', it can be reasonably said that the environment primarily dictates the overall 'dance', the choreography, you might say. So, for example, an atom responds to its molecular surroundings largely according to the influences of that environment, which reflect the presence and status of any and all other atoms in the vicinity, such as electric charge or electromagnetic energy, and gravity. But we can also see that in living processes, molecular organisation is governed by and within, an environment more complex still, which directs its behaviour - i.e., the organic domain. If we were to examine and compare molecular organisation from animate and inanimate sources using suitable equipment, such as an electron microscope, the only difference that could be found at that level of observation would be in the degree of complexity present, no extra material ingredient, or unique form of influence, that could be identifiable as being present only in living forms. Nevertheless, the rules of living organisation direct the behaviour of the otherwise inanimate molecular chemistry. Simply put, living processes govern the organisation of inanimate molecular structure to create all the varieties of organic form such as proteins and all the internal and external features of cellular organisation. The central instrument of organisation within that environment is the genetic code contained within the cellular DNA - itself a molecular structure, but the wider environment this reflects is that of ecology, in which the main influences of selection pressure are expressed (at the risk of stretching the analogy just a bit too far, you might consider the function of genes as that of choreographer, directing the molecular dance, according to the music provided by the ecological environment...).

Without dwelling on fine detail for now, this illustrates in a general way the *hierarchy of relationship* present and expressed through evolution. This relationship can be seen at all levels of observation; the molecular environment governs the behaviour of atoms, which effectively includes that of sub-atomic particles, the organisation of organic life governs the molecular domain, while higher up the chain of hierarchy, conscious experience has a dominant influence in the shaping, via behaviour, of organic life. In each case, the higher domain is seen to direct the behaviour of its supporting domains; in physical terms, biochemistry governs the behaviour of the supporting domains, molecular and atomic, all the way down to the level of sub-atomic particles. Nevertheless, it is a mistake to think of this hierarchical relationship only in physical terms, it concerns the *development of complexity* rather than the material of the content. It is against this background that phenomena such as conscious cognition and intelligence have evolved. In strictly physical terms these may seem insubstantial and secondary, but in terms of evolutionary organisation their influence is no less substantial than that of the molecules of which living form is composed, and in terms of hierarchy, they represent the advance line of evolutionary development.

To help get a grasp of this idea in contemporary terms we could say that evolution is a principle that concerns complexity rather than material content, you might say, of 'information' as distinct from molecular structure per se. DNA itself is a high-level physical expression of this principle; the domain of conscious experience is a further level. As DNA can be seen to coordinate the organisation of molecular structure - including its own - consciousness organises its contributory organic structures, including the DNA (indirectly, via the normal processes of natural selection). Thus, have structures including the nervous system and the brain itself developed, or *emerged*, under the auspices of conscious experience, and the evolution of complexity continues in this non-physical sphere, supported by all the prior domains of organic and inorganic nature.

Summary

Viewed in terms of self-organising complexity, evolutionary development can be seen as being, as-it-were, layered, where the lower levels of these layers provide the content and structure that makes the higher levels possible, while the higher levels direct the behaviour patterns of the lower. Accordingly, the higher levels are effectively dominant, the lower levels supporting, or subsidiary. Although we may perceive and describe these various levels as appearing to obey different 'laws of nature', their development is continuous, unbroken, and hierarchically structured. From this perspective, principles of evolution should be viewed as being represented equally throughout all domains, physical, organic, and conscious (including constructs of thought). In terms of organisation, while the laws and principles of the sub-atomic and molecular domains are fundamental and provide the building materials of the universe, their organisation is directed from the higher levels of complexity.

Chapter 4

Natural Selection and Ecology

Outline

The principle of natural selection is central to the Darwin/Wallace theory of evolution, summarised in the well-known term 'survival of the fittest'. This expression wasn't coined by Darwin or Wallace, but by Herbert Spencer after reading Darwin's book, 'On the Origin of Species'. Darwin liked the term and thereafter adopted it for general use (Spencer was an economist and already used the term in that context). Both expressions, 'natural selection' and 'survival of the fittest' mean essentially the same thing, which is based on certain observations made by Darwin and Wallace. A summary of this is that in any given situation, some creatures, on account of their characteristics, will be able to survive and propagate more successfully than others. Inheriting these characteristics, their offspring will also tend to be more successful and prolific than others in similar circumstances, resulting in the perpetuation of those features within their species. The circumstances involved include many different aspects, from basic considerations such as geographic or atmospheric conditions - temperature and moisture for example - to those of available food sources and potential predators, in other words the organic context. This chapter briefly outlines the main ideas involved, extending them in some cases to broader, inanimate contexts, that I have touched on, with some additional observations of my own.

Having discussed in outline the basic idea of evolution, and introduced that of evolution as a universal phenomenon, in this chapter I'll now consider a little further some concepts that are well-established features of modern evolution theory, extending these where I feel it appropriate.

Natural Selection

Central to these concepts is the idea of 'natural selection', which is based on a recognition, of long standing, that particular forms of species, whether animal or vegetable, can be chosen and preferentially cultivated; in other words, *selected*. Farmers and animal breeders of all kinds from all cultures have long been aware and made use of this. The insight that Darwin and Wallace captured and investigated in detail, was that nature does this all the time. Simply, what works best survives, prospers, and propagates. Viz the term, 'natural selection'.

Variation and Adaptation

Darwin observed that the way in which 'selection' happens naturally, is that very slight differences in form occur between otherwise identical creatures in any given situation, which are then inherited by subsequent generations. Changes that result in improved survival and propagation are referred to as adaptations. The cumulative effects of such slight changes over great periods of time have resulted in the formation of every physical feature to be found in living nature, from eyes to wings to legs, the spinal column, skin, blood, the detailed structures of the very cells that make up every living thing - to name just a little of the vast repertoire of organic invention. Negative variations tend to reduce or eliminate the reproductive line concerned. This overall distinction is usually referred to in terms of 'competition' because the 'best suited' or 'fittest' are those which survive and propagate most effectively, and the overall result is referred to as 'natural selection'.

A second observation is that the source of variation that underpins this process appears to operate randomly. With very detailed observations, Darwin noted the expression of such variation and its effect on the development of species but had little detailed knowledge of the mechanisms of inheritance involved. (Perception of heredity had a long history in everyday observation, particularly farming practice, and more detailed inquiry, but was brought into focus in a scientific way by Gregor Mendel in 1865, although his work wasn't really recognised until the early 20th century).

It was only much later as understanding of the functions of DNA grew that the understanding of genetic mutation was fully introduced and became part of the parlance of evolution theory. Nucleic acids, which are the basis of DNA, were first identified by Freidrich Miescher in 1869, but their molecular structures and the nature of their contribution to inheritance were only unravelled in 1953 by Watson, Crick, Franklin and Wilkins.

A very basic summary of the principle of natural selection then is this; that which works best tends to stick around and become established; that which doesn't tends to disappear. Although this may sound like a rough principle, in practice all the incredibly fine detail of organic nature results from this process, minute variations contributing towards major changes over time and ultimately, the shaping of all complex form. The analogy comes to mind of pixels in an image. Each pixel alone may appear insignificant, but together an image is expressed, and the finer the scale of the pixels the more detailed the image.

Reproduction

A key point of distinction between inanimate and living nature is that of reproduction. This may seem obvious and intrinsic, yet this observation derives from a time when the workings of neither sphere was much understood and no direct links could be perceived between that of inanimate chemistry and organic processes. Later, when the function of genes was deciphered, reproduction became intimately associated with genes, DNA, and the complex structures of cellular organisation.

It's a good idea to take a close look at what we mean by reproduction. The general meaning of reproduction is the generation of copies of an existing form. This concept is generally associated with cellular life-forms, but it can be applied more broadly, at any level of observation, from that of molecular organisation to the creations of conscious culture. The big distinction really arises only when we think of it in terms of complex coded information, as represented in the genetic code within the living cell, but even inanimate molecular organisation frequently involves the regular repetition of patterns of structure of many different kinds. Repetition at this level could easily be considered as a form of reproduction, in the sense of the continuance and maintenance of structural form. These forms derive from the innate inclinations that the atoms of elements carry on account of their structure, in particular, their valence, which predisposes the nature of their relationship to other atoms. This is the basis of all molecular structures and accordingly, of all mineral chemistry. In any given context, living or inanimate, individual atoms behave according to the influences within their environment, constantly seeking a state of maximum balance of charge between their central nucleus and the electrons that surround them resulting in molecular bonding of various types. While the nucleus largely represents the core 'identity' of the atom as an element, electrons are relatively free to behave and be shared between atoms on a much more flexible basis. This flexibility is the practical basis of valence and largely underlies the patterns of relationship that are represented throughout the vast range of compounds and minerals that form, exist, change, and exchange over many different scales of time.

This formation, change and exchange is intimate and dynamic, but central to it is the constant search for equilibrium. Under the influence of conditions prevailing in the immediate environment atoms shift or adjust their allegiance. Within this dynamic, naturally some arrangements will be more stable than others, and such arrangements then contribute significant patterns of influence within the environment. At this level then, the tendency for more stable patterns of organisation to become established represents the heart of reproduction, not so much in terms of individual, discrete, entities such as cells, but as a still more basic principle of molecular organisation. Here, the idea of reproduction relates more generally to structural patterns of relationship. This tendency, at the molecular level, is ever present as the cementing principle that underlies the much more complex form of reproduction that we observe at the level of cellular life.

Reproduction therefore doesn't suddenly appear as a unique feature of life. In a much more open sense, it is a feature of molecular existence and beyond that, is expressed in one way or another in every domain of evolution. Having made that general point though, clearly cellular life depends on the reliable reproduction of molecular structures of a degree of complexity never found without the contribution made by genetic coding. The question then arises as to how could 'reproduction' at the molecular level have developed, or evolved, to this very much more refined and defined form that we find in organic life? We'll consider this now, here and in the next chapter.

Variation Through Mutation of Genes

Simply put, the term mutation refers to the phenomenon of spontaneous and random change in form. In the context of organic evolution this refers specifically to inheritable molecular changes to the structure of the DNA genetic code. Broadly speaking, mutation arises because of an entirely normal inherent feature of molecular organisation. As all the constructs of organic life are molecular, any part of those structures is susceptible to disturbance that may be introduced in the normal course of organisation at the atomic level. For example, and particularly, the constant bombardment of our planet by a variety of high energy particles from the sun and other cosmic sources is a major source of disturbance to the integrity of complex molecular structures. The atmosphere of Earth provides a considerable degree of protection from this bombardment, but not complete protection. Within organisms this disturbance can and does take place at all levels of structure, but is most significant evolution-wise, where reproductive DNA is affected, because such alterations are passed on to the next generation, with consequences that are integral to the way in which evolution proceeds. Other important sources contribute to the variations which emerge, not least the sexual exchange of genetic information, but we may just focus a little further for the moment on mutation.

The genetic record is based on a complex molecular structure, the integrity of which depends on the stability of its component parts (as do all the processes that take place within the cell). When we talk of mutation in the context of evolution, although its effects may be observable at the level of the expressed form of the organism, the introduction of random variation is at the level of this core structure, as it underlies the nature of the information stored. This genetic archive is constructed with four elements of DNA code (or RNA in the case of many viruses), each of which is essentially an inanimate molecular structure. While such structures are relatively stable - and the integrity of all life depends on that stability - they are not absolutely so; like everything else in nature they can be corrupted by numerous kinds of influence. Typical sources of disturbance to their structural integrity are strong sources of electromagnetic energy, ultra-violet light, x rays and atomic radiation, to give a few examples. But even if one could create ideal conditions as far as possible, certainty regarding the stability of molecular structures could still never be entirely guaranteed. In any case, if molecular integrity were to be inviolably stable, there could be no variation and therefore no evolution. Mutation is essential to life...

In addition, the position of such disturbance is important. When it occurs during the life of an organism it can result in damage ranging from minor effects to catastrophic disease such as cancer, but mostly that damage is limited to the organism in question. The most crucial areas concern the cells that participate directly in reproduction. Changes/mutations here have a high likelihood of being passed on through inheritance. This would apply whether we were considering the division of the simplest of microorganisms, or the sperm and ova at the level of more complex life forms. Concerning the contribution made by mutation to evolutionary developments, there is a significant difference between the two ends of the spectrum of organic complexity, i.e., simple single cell organisms and complex multi-cellular organisms. This difference is that the more complex the organism is the less likely mutation is to provide 'useful' variations. The simpler the organism, the greater is the possibility that single mutation events may produce changes that significantly benefit its prospects of survival and propagation. The reason for this should become clearer from the following discussion.

Variation Through Exchange of Genes

There is something of a widespread impression that mutation is the only source of genetic variation. This is a bit of a misunderstanding, or oversimplification. Mutation is clearly the bedrock source of variation at all levels of cellular evolution, but the direct transfer of genes between single cell micro-organisms, and sexual engagement between multi-cellular organisms, introduce another important level of variability, based on the exchange of genetic information of a much more coherently structured form and in more controlled ways. The basis of this exchange still involves a high degree of randomness, but largely, the genetic structures and information involved in these types of exchange is complex and has been established, on both sides of the exchange, for many generations, as distinct from random mutation which is by definition, entirely unstructured. Genetic mutation remains ever present as a contributing factor to the outcome of reproduction, but at higher levels of complexity it mostly makes only relatively minor alterations to established genetic architecture. These differences are still important sources of fine variation with sometimes significant consequences, mostly negative in the context of multi-cellular organisms, but are still also capable of creating advantage occasionally at that level. The variations, which Darwin and Wallace studied, would have arisen from the conjunction of these two sources, mutation, and sexual blending. An analogy could be made here with the contents of a pack of cards in the making; if we regarded the details on each card - the numbers, images and shapes - as the result of historic successful mutations established within the genetic code, the shuffling of 'complete cards' of genetic information by each occasion of sexual exchange provides the randomising of their distribution at a more complex level without disturbing their integrity. (Though this only begins to touch on the complexity involved here, because, to extend the analogy just a little, the images and numbers on those cards would also represent many layers of prior organisation arrived at cumulatively in essentially the same way over countless generations of reproduction and exchange over hundreds of millions of years. An important implication of this is that the direct contribution made by mutation would be proportionally greater in the evolutionary distant past, decreasing with the increase in complexity over the passage of time.)

While there is a vast range of directions and forms that life has taken, as far as complexity is concerned there are three types of cell structure which are well represented on our planet, and the degree and manner of variation that occurs in their domains have some important differences. The two more elementary of these cell types, are, for the most part, represented by single cell organisms known as bacteria and archaea, all of which are known as *prokaryotes*, which essentially means that they don't possess an isolated genetic nucleus, or mitochondria, and are self-regenerating, without the benefit of sexual exchange of DNA (although there is some exchange of genetic material between individual cells of this kind, and some complex exchange chemistry). Random genetic variation on this basis, mutation, figures highly in their evolution. Because of their relatively simple structure, slight molecular variation may make very significant alterations to their success or failure in propagating within any given environment.

The third type, represented by all complex multicellular life forms we know of, and certain single cell types, such as the amoeba, is known as *eukaryotic*. The eukaryotic cell contains various membrane isolated components, or *organelles*, in particular, the nucleus and genetically independent ones called mitochondria that provide energy for cell metabolism. The mitochondrion is a kind of ancestral prokaryotic lodger, with its own genetic history, that initially took up its residence within that cell type, possibly 1.5 - 2 billion years ago. Another genetically independent feature that emerged in eukaryotes had its origins earlier in prokaryotes, where it created the condition of free atmospheric oxygen the mitochondrion exploits. This is photosynthesis. Combined with the emergence of distinct, organelle membrane 'packaging', these features laid the ground for the developments of higher complexity life.

Sexually based reproduction, as we know it, is unique to complex - *eukaryotic* - life forms, because it represents the exchange/mixing of complete units of complex genetic information that don't exist at the level of the simpler - *prokaryotic* - forms. Genetically, the entire function of sexual exchange, its raison d'être, is to facilitate the exchange of complete systems of tried and tested architecture between individual members of a species, to their possible mutual advantage propagation-wise. This form of genetic variation still employs a strong element of randomness but effected in a much more controlled way that maintains the integrity of complex structures of cellular organisation that have been assembled over many, many generations.

As touched on, in the prokaryotic domains, there is also what is known as 'horizontal' transmission of genetic material between independent cells, and some complex exchange of chemistry, but to a much less structured degree than that of the sexual exchange of the eukaryotic form. Nevertheless, even at that simpler level such exchange is by no means entirely random, involving the exchange of tried and tested structural features. While mutation provides variation at the most minute scale of evolutionary process, sexual and horizontal exchange provide it at a more macroscopic level, facilitating variation at much higher levels of complexity than genetic mutation alone ever could. If genetic mutation could be represented as grains of sand, sexual exchange would be represented by the exchange of bricks composed of such sand, that have been fired, tried and tested over countless generations, and of complete elements of architecture composed of such bricks, from marble flooring tiles to slate roof tiles, from doors and windows to complete buildings including all the wiring and plumbing; and with all varieties of camouflage and decoration...

Ecology

The term ecology is used in reference to the complex network of relationships in which organisms participate and represents the context in which natural selection takes place. The most obvious background features of that relationship are the inanimate physical circumstances prevailing in any given place - the geography, climate, altitude etc. These of course are variable but provide a common context for all the inhabitants. But more importantly as far as understanding the meaning of ecology is concerned is the fact that most organisms are only able to survive and propagate by consuming material provided by other organisms. This is commonly referred to as the 'food chain' or 'food web'. Essentially, this means the consumption of organisms by other organisms. By this means, the sustenance required to support the metabolism of more complex organisms is generated mostly at the level of simpler organisms, while *their* sustenance comes from their intake of simpler organisms still. At the lowest level of such a chain are organisms known as autotrophs, which can metabolise material such as carbon and nitrogen directly from their surroundings and without which, such a chain would be unable to sustain. Cyanobacteria, for example, have been present in the biology of our planet for at least two to two and a half billion years, possibly much longer, and represent an important stage in the ways in which life has developed. As well as drawing essential elements from inorganic surroundings, and therefore contributing to the foundational linkage between the inanimate domain and life, they appear to be a major contributor in the evolution of photosynthesis, as we know it, which has resulted in the presence of free oxygen, the source of chemical energy on which most complex life depends.

An important part of this overall process is the participation of organisms that decompose dead organic material down to - still complex - compounds that can be absorbed easily by other active forms. Another aspect is the involvement of energy sources that drive the biochemistry of all these processes. To all intents and purposes, for billions of years now, a principle driving energy source for life on Earth has been, and is, the sun, both directly and via the biochemistry of photosynthesis. Before the evolution of photosynthesis, the sun would still have been a major energy source, but there are good reasons to suppose that very early forms of life may have been sustained primarily by the molecular energy contained within complex mineral structures and in particular, by volcanic thermal vents deep in the sea.

In practice then, ecology directly reflects the hierarchical nature of self-organising complexity, i.e., evolution, and the essential continuity of the support systems it generates and depends on, with its strength but at the same time inherent fragility. Every link in the chain supports the next, in every direction, and contributes to the shaping of the dynamic status quo in any given context. The ecology of land, sea and air have their own features which are distinct, but of course have common ancestral roots and are intimately interwoven.

A central principle of ecology is *balance*. Everything, at every scale, has influence on everything with which it is interconnected, directly 'above' and 'below', and indirectly - laterally - on other populations. Populations of any species at any scale are supported and constrained by the nature of the species that surround them. Principally those they consume and those by which they are consumed, but at all levels this also involves complex symbiotic relationships which are integral to that balance, such as those between fungi and plants, by means of which both derive nourishment. Again, this principle of balanced relationship between life forms reflects the same ubiquitous principle to be observed throughout all levels of evolutionary complexity, inanimate or animate, which I've discussed in the chapter on self-organisation - that of dynamic equilibrium.

Competition and Cooperation

The active principle of natural selection, largely regarded as central in the view of current convention, is conceived of in terms of 'competition'. For sure the evolution of complexity at any level rests on the comparative success of particular forms, but 'competition' can only take place between structures that already exist, and at any level of complexity, that involves some form of 'co-operation'. It is now generally recognised that 'co-operation' is an essential partner to 'competition' in the shaping of form within all domains of evolution, and the evolution of complexity perceived more widely throughout nature rests on the relationship between these two features, which, again, is dynamic. I should say here that the idea of co-operation in the context of evolutionary theory has been, and still is for many, anathema. It's not that long since a prominent expression one may be likely to encounter in reading or in discussion of evolution was that of the 'selfish gene', with a more-or-less complete emphasis on the aspect of competition. As I understand it this term was intended as a metaphor to help focus on the central role of the gene in evolutionary processes in the organic domain. This is no doubt an effective approach, but new insights have been emerging that greatly expand the perspective of evolutionary processes beyond focus on the demands of the gene per se.

In everyday circumstances the term 'competition' generally suggests deliberate behaviour, but in the context of evolution it has principally one extremely wide-ranging meaning, which is implied in the expression 'survival of the fittest'. In any given set of circumstances where there are limited resources and particular environmental conditions, those creatures best equipped to survive and propagate become predominant. This is the meaning and use of the term 'competition' in this context, and doesn't refer to intentional, or overt, competition per se. Similarly, 'co-operation' here is not intended to suggest deliberate action but is simply a general term representing situations where selective advantage is gained from some degree of mutually beneficial interaction. Of course, both competition and co-operation do find overt expression in animate nature at many levels and in many ways but in general these terms are a short-hand way of summing up complex and subtle processes involving every imaginable parameter of circumstance, from weather and geographical conditions, to prevailing biology and ecology. In this sense then neither term, competition, or co-operation, implies conscious action or intent per-se but is merely a means of characterising the relationship of the participants, from the point of view of an observer, and attributing these qualities to them. In the context of evolution therefore these general descriptions can be used in relation to the reproductive success of living processes at all times in the history of genetic reproduction from its origins to the present day.

As with the term 'reproduction', a more subtle view into the meaning of the ideas of 'competition' and 'co-operation' can be found by considering evolutionary processes at the molecular and mineralogical levels. In Darwin's time little or nothing was known about the generation of elements by stellar evolution. The concept of 'natural selection' was perceived to apply strictly to the domains of organic life, but a great deal more has been discovered since then that broadens perception of certain aspects of this principle as being represented more widely. As a general observation, the principle of 'selection' can be applied not only to organic life forms but further, to all kinds of situations of self-organising complexity, from the formation of atoms, molecules, chemical compounds, and minerals, through the organic domain, to those of social/cultural interaction, across all traditional boundaries of definition.

It is only in recent decades really that the degree of complexity involved in the formation of planetary minerals has begun to be fully recognised. I am not going to try to go into this topic in any detail, as it is a vast subject of inquiry. For my purpose here I just want to make the observation that all minerals consist of molecular combinations of elements and that these combinations represent relatively stable forms that arise and persist in the circumstances to which their component ingredients are subjected at the time and place of their formation. These circumstances are many and varied, often highly energetic and have occurred over great periods of time. Our planet Earth originated through the gravitational gathering of dust and fragments, largely arising from the dispersion of remnants of generations of earlier stars, such as nova or super-nova and other more gradual events of stellar dissolution, coalescing along with the other planets in our solar system, around what was simultaneously becoming our sun, itself arising in a similar fashion. In the gathering, these fragments have been melted by solar energy, combined, and re-combined many times, then subjected by the heat and pressure of the Earth's tectonic processes to continual cycles of re-combination.

At all stages in that processing and in all situations, the most stable molecular structures and forms are the ones that become established, laying down the substrates for the developments that follow in 'preference' or superiority over less stable structures. Always there are balances in the possibilities of chemistry involved in these combinations, and at that level, particularly of fine balances, the direction of chemistry is governed by the predominant mineral and molecular environment. Molecular structures that develop in this inanimate context include many we refer to as organic structures (although these don't originate from living processes; they are referred to in this way because their chemical form is intimately associated with molecular building blocks that organic processes employ) and can be quite complex and hugely varied. From the perspective of molecular chemistry, we may regard many of these structures as fragile, nevertheless, the possibilities presented by the molecular environment determine what can develop and be considered 'stable' at any given point. The context is quite analogous to that of organic ecology (albeit, and importantly, without the feature of self-contained reproduction) and this description outlines the perception of 'selection' as an inherent feature of the processes involved. Clearly though, in this inanimate domain the metaphors of 'competition' and 'co-operation' feel inappropriate and more difficult to apply - highlighting their anthropomorphic background - and yet essentially a similar kind of dynamic is recognisably at work in molecular organisation. At this level, 'co-operation' represents molecular coherence arising spontaneously from innate principles of collective atomic engagement, while 'competition' reflects the relative stability of molecular structures within the context of given environments, and the consequent tendency of certain combinations to 'prevail' over others. Both of these factors we understand in terms of physics and chemistry, but essentially the same basic principles of selection are represented here as in the organic domain. In effect, the terms co-operation and competition represent two sides of one coin. The image we perceive depends on which side we are paying most attention to, but the other side is always present and implicit. As ever, the perspective of the observer directs the gaze.

It's also apparent that all higher complexity life arose in the first place through a variety of forms of genetic exchange or symbiosis between otherwise distinct prokaryotic micro-organisms, later in much more organised ways such as sexual exchange, later still in the kind of collective, 'social' co-operation seen in many types of colonies of insects such as bees, ants, termites as well as that of many other animals. Clearly this continues to be the case at all such levels of organisation, including of course human society. So, we need to consider both notions, competition, and co-operation, and how they relate. One way of grasping this view, in contrast with the emphasis on competition, is to consider that the prime 'purpose' of life *overall*, is to propagate by all means that are effective. From this perspective, random mutation underpins only one aspect of the partnership between competition and 'co-operation' in the emergence of organic complexity, and as I've discussed, is most significant at the lower rungs of the ladder. Above that level, competition takes place between all the range and variety of complex forms of organisation - i.e., co-operation - that life, in its ingenuity, can contrive in its efforts at propagation.

Complexity

In recent decades it has become increasingly clear that the rules of nature that govern events at the molecular level are indistinguishable from those of living process, except in terms of the complexity involved. For that reason, recognition of the development of coherent complexity is a basic requirement of any understanding of evolution and nature as a whole. (As distinct from chaotic complexity. A comparison might be made here between, say, a ball of wool and a tangle. Both may be perceived to be 'complex', but one as coherent, the other as chaotic.)

Inanimate chemistry rests on physical rules of organization and behaviour that we have to a large extent come to understand. However, coherent as these rules are, they give rise to a great deal of freedom of possibility for the directions chemistry can take at any given point of circumstance. What any atom will do at any time is determined by a combination of its internal nature and the conditions prevailing in its surroundings. Internal conditions may be a given perhaps, but the environment can take a vast range of possible circumstances. This field of influence is so open and energetic that it makes it difficult for organised complexity to sustain more elaborate forms of structure that may begin to form in the short term. Most molecular formations found in life have their roots in simpler forms widely found in inanimate nature, but in living organisms these forms have developed into highly complex structures that could never have formed or sustained under usual inanimate conditions. Even the simplest living entities are still highly complex (coherent) structures, and fragile under the intense and irregular pressures of open chemistry. Coherent molecular organisation constantly emerges in inanimate chemistry, but in free natural circumstances outside of laboratories this is unable to develop beyond a relatively low degree of complexity, though in wide variety. Many of these molecular forms are present and available as building materials of organic life, but are no more organised beyond that than say, sand is, compared to bricks. By this comparison, the levels of organisation achieved by life in this respect could be compared to the construction of vast cities from the raw material of sand.

Regarding our understanding of the transition of evolutionary complexity through molecular self-organization, from that of inanimate chemistry to that of living processes; this is a significant area of interest and inquiry, with many theories and avenues of inquiry currently being explored and described. Exploring these in any detail is beyond the scope of this essay but there is a particular line of thinking that interests me, which I'll discuss here.

Naturally, in examining the evolution of life the focus of inquiry tends to be aimed towards the origins of genetic reproduction. The evidence is now very strong that all life as it exists on this planet can be traced back to one common original cell, sometimes referred to as LUCA (last universal common ancestor). Although this is as far back as we can deduce currently with a reasonable degree of certainty, this ancestral cell would already have incorporated many basic component structures that have continued to be represented in all forms of life on earth ever since. These structures, including that of the genetic DNA, could not have appeared suddenly overnight however but must have been the result of much prior 'experimentation', in the sense of molecular 'co-operation' and 'competition'. It is difficult to conceive of circumstances in which a highly complex but fragile structure like DNA could have evolved without - at least - the protection provided by the outer cell membrane, but how could that membrane have evolved without the instruction of the DNA...?

Regarding this question, apart from core structures such as DNA and/or RNA, the outer membrane enclosing the cell contents is an essential common feature of all known cell structure that may well predate the emergence of LUCA per se; could it also pre-date the development of DNA/RNA? This membrane has developed greatly in variety and complexity over the several billion years for which it has been participating in the evolution of life. Nevertheless, for as far back as we can reasonably deduce, its molecular structure has been based on that of lipids, which also occur widely in inanimate nature. I won't attempt to go into detail here but basically, the significance of lipid chemistry concerns the behaviour of certain compounds, known as phospholipids, in relation to water. Commonly, we know of lipids as fats and oils of which there are many different kinds, phospholipids represent just one branch of the varieties that exist. In terms of chemistry, the feature of interest is the fact that the arrangement of the phospholipid molecule is such that it is polarised in relation to water. Accordingly, one end of the molecule is attracted to water, the other end is repelled. One consequence of this that occurs widely in inanimate nature is a tendency in certain circumstances for these molecules to spontaneously take on a double layered spherical form in the presence of water. These are referred to as vesicles, meaning tiny vessels. The structure of these vesicles is such that they contain water internally while resisting it externally, and they are quite capable of randomly enclosing all sorts of additional molecules as they form. Bearing in mind the ability of water to dissolve and contain a vast range of substances to some degree, it is virtually inevitable that this would happen. Any chemistry that may take place within, or become incidentally incorporated into, these structures is therefore naturally partially isolated from the surrounding environment. Most such chemistry may be of no consequence, nevertheless there is good reason to suppose that this may have provided ideal conditions for further developments in complexity to proceed, leading to the emergence of primordial cellular structures capable of some limited degree of complex molecular reproduction, and ultimately to the appearance of RNA and DNA, whose structures are at the heart of all known life since the appearance of LUCA. It should also be understood that along with many other complex molecules, lipids are a feature of (inanimate) cosmic chemistry and have been present widely on earth since its formation. Accordingly, the spontaneous self-organisation of lipids in the presence of liquid water probably began very early in the formation of the planet.

However, regarding this view of the possible role of lipids in the origins of life, this is only one of many theories and related considerations being explored. Bearing in mind that as it would have contained all the essential components of life as we now know it, including DNA, RNA and essential mechanisms of metabolism, LUCA already represented a very advanced stage of molecular evolution. Huge gaps remain in our understanding of the transition from inanimate chemistry to the emergence of LUCA... Although all known cellular membrane chemistry is based on that of lipids, there are other ways in which localised isolation arises spontaneously. It is possible that partial isolation originated through some other means and that lipid chemistry came later. Such possibilities are being explored. It may never be possible to be certain whether lipid chemistry, or some other form was initially responsible for the emergence of the cell membrane, but perhaps more significant for insight into evolutionary processes is the idea of partial isolation.

'Competition for resources' is a directing factor at the level of molecular organisation just as important as it is at that of organic life. From the perspective of complex development, in addition to providing a protective environment for nature to 'experiment' with delicate molecular structures, partial isolation provides a situation for 'competition' to take place between such developing structures. Represented in different forms, from the level of molecular organisation, through all domains of evolution, partial isolation appears as an intrinsic feature of complex development.

Emergence

Emergence is a term that has come into use, chiefly in the context of organic processes but also more widely now, to represent the principle whereby new complex forms and domains of phenomena develop from more elementary backgrounds of processes and events. One major example of this is that, for more elementary single cell organisms, widely represented on earth, the entire contents of the cell are contained within a single membrane structure, while more complex organisms possess a cell structure in which certain component parts, referred to as organelles, are (partially) isolated by individual membranes from the rest of the cell apparatus. A crucial feature of this development, that makes it possible in the first place, is the presence of organelles known as mitochondria which provide energy for all of the cell's energy needs. These have their own independent DNA and genetic history. This arrangement appears to be essentially symbiotic in origin, with roots in ancestral events that represent a transition from the simpler kind of cell structure - which continues to be represented widely on the planet - to a considerably more complex type. The simpler cell structure is known as prokaryotic the more complex as eukaryotic. Simpler cell types do on occasion give rise to collective behaviours superficially similar in appearance to the complex type, but all true multicellular organisms appear to be based on the eukaryotic cell type. This transition from prokaryotic to eukaryotic cell types is a classic example of emergence, where one organised form gives rise to a new one, with consequences to directions taken in the ongoing developments of form. In this example, the shift is great because the entire range and direction of possible developments has changed; nevertheless at the same time, many core characteristics of the cell's component structures remain. It can be reasonably said that emergence of this sort is both sudden and gradual. The presence of mitochondrial DNA in the eukaryotic cell suggests that at some point in the life cycle of one cell, back in the interaction between simple cells, one microorganism absorbed another, with the result that it became the first in the line of those subsequent microorganisms whose reproduction was to produce all the complex multicellular organisms that now inhabit the planet. This means virtually everything from fungi, through plants, to all the animals and insects of the air, sea, and land, including ourselves. Quite a result for one cell amongst countless trillions, but this process of cellular 'invention' is repeated continuously, if usually less obviously everywhere throughout organic nature - every variation, adaptation and species resulting from the expression of this same principle, emergence, constantly at work, frequently arising through symbiosis - cooperation - in the first instance, rather than competition per se (photosynthesis is another significant earlier example of this).

The transition from inanimate molecular processes to organic biochemistry can also be understood in terms of emergence (and this is an area of intensive experimental exploration in evolutionary biology). Meanwhile, cognition amongst animate life forms, clearly seems to be emergent from let's say, a more 'vegetative' background, though as far as I am aware, this is as yet, much less explored.

Emergence as a principle then, is most often identified and associated with natural selection, ecology and living processes, but more generally it applies to all dynamic complex systems, animate and inanimate. For example, at an early stage in cosmic evolution, after the elements of hydrogen and helium had taken form, but before the appearance of the first stars, there would not have been much for a visitor from another universe to see, no stars or galaxies, no planets, comets or even rocks. Possibly light, but nothing yet to see... But once cosmic forces had drawn those first atoms together to produce the first generation of stars after some millions of years (timing uncertain, but possibly between 50 million and 200 million years, based on current theories, galaxies also began to form, and the universe as we know it now began to take shape. This transition from widely dispersed atoms to the formation of stars is an example of emergence in the inanimate domain - not the first, although perhaps one of the first that we can begin to conceive of visually - nor the last, since the stellar manufacture of further elements beyond hydrogen and helium represents the ongoing process of emergence on a cosmic scale.

'Emergence' should be understood to be a continual and ongoing feature of evolutionary process. Another word for it is 'transformation', which suggests of course, the change of 'form', but which also carries the implication that the underlying 'substance' doesn't change. In fact, change in form in the sense of coherent complexity depends on the reliable consistency of under-lying 'substance'. The entire range of studies in which the sciences are engaged could be described in such terms; the search to dis-tinguish 'substance' from form, to find 'essential' ingredients/principles and to understand the forms that emerge from or through, them.

Another important example of emergence is that of con-sciousness from organic roots. While the transition from inanimate nature to organic processes is clearly a very considera-ble one it can still be interpreted in terms of molecular complex-ity, but consciousness is not a 'physical' phenomenon, it clearly exists, as real and consequential in its own right as iron or air, and that existence rests on a very tangible basis of biological and molecular structure; but it can't be weighed or measured, put in a jar or have its temperature taken, nor can it be inspected through a microscope or probed with lasers. The point here concerns the limits of our models of observation and conception though, rather than anything more mysterious. This transition, and the relation-ships of the factors involved, can only be understood in terms of evolutionary complexity rather than in terms of 'material' per se. All the stages of transition in emergence are necessarily linked, but clearly, the form each stage takes may be very different from that of prior supporting levels, and not necessarily definable in terms that translate directly from those prior stages. For that reason, the terms of description we can apply in inquiry into the behaviour of atoms and molecules at an inanimate level don't suffice to provide full understanding of the organic sphere, and the terms of description we use in that context don't suffice to provide full understanding of the domains of conscious experi-ence and resulting behaviour. One thing we can say with reasonable certainty is that there is no reason to suppose that the principle of emergence is no longer represented in the conscious domains of evolution. That seems very unlikely.

From the perspective of imaginary observers located within each of these domains of nature, higher levels of organisation may well be imperceptible, perhaps even inconceivable, while the underlying levels, may be more easily recognised and investigated. If you were to imagine for a moment, being an atom, living in the world of atoms, and aware of your environment, you may understand all your duties and functions as an atom, but with no, or little and vague, awareness of the world of complex molecules, comparable perhaps to a soldier in a large army. Further, you would have no knowledge whatever of influences at the organic level that govern those molecular processes. These different levels of complexity could be considered spheres of hierarchy. We can say that the higher sphere governs the behaviour and organisation of the lower. This isn't to suggest that any level is superior to others; all levels are simultaneously essential and function in a completely integrated way. Again, this represents the idea of *hierarchical relationship*, which also appears in many other contexts that may not be immediately apparent but is a key to understanding how evolution operates.

An important point here is that from a conventional perspective, cause and effect at the microcosmic level is considered to drive evolution. This is commonly read as implying that living dynamics, and the emergence of consciousness and intelligence, are simply extensions of the same 'mechanical' processes of nature as are identified at the inanimate level and should be understood on that basis. However, laws of nature as we have interpreted and come to understand them are derived from the observation of events in terms of views/models which also evolve as we learn more. This is exactly how science works - events are studied and principles of relationship are deduced by analysing those events against the background of prevailing knowledge. As we do so new dimensions of understanding open and alter our frames of reference. In short, the paradigms of interpretation also evolve.

Laws of physics are studied and identified at a fundamental level to find understanding of universal organisation. Quantum theory apart, causation tends to be interpreted as particular and linear, but in no domain of nature does causality operate in a strictly linear way. As I have discussed, organisation is a result of the interaction between what could be thought of as 'internal' and 'external' factors, or the microcosm and macrocosm (and this is always hierarchical; the immediate circumstances represent the 'microcosm', but always reflect the greater environment to which they are subject), or you may say, between individual 'components' and their environment, and can't be perceived in its dynamic variety and complexity purely by studying smaller and smaller components and their rules of behaviour - although knowledge of how they function is important. That's perhaps like trying to comprehend a TV image by investigating the details of the pixels, and then the internal structure of the pixels and so on, and the electronics governing them. Perhaps somewhat paradoxically, viewed in terms of hierarchy, the image being generated dictates the behaviour, or presentation, of the 'pixels' rather than the other way round. It might be more appropriate therefore to think of the understanding of the 'macro' and the 'micro' scales of observation as complementary, both being essential to obtaining a complete view of the workings of nature.

Regarding this point, the basic principles of the classical view of physical processes - which to a large extent still prevails in our attitude to the physical sciences - were established in times when the context of observation did not include any perception or understanding of the evolutionary dimensions of the dynamics involved. The range of the scale of observation was strictly limited to that of the dynamics of physical processes that could be perceived, explored, and understood - without that very important feature. The context has now changed considerably, and some recognition of the evolutionary dimensions of processes is now a significant feature of all scientific inquiry. This is partly because the classical approach of physics theory has been revolutionised internally you might say, by the introduction of relativity and quantum theories, both of which have reduced the grip of linear views of causality. But it is also because the growing recognition of evolutionary process at every level and stage of the development of complexity is beginning to alter the framing of our perspective on the relationship between natural principles at every level of nature. It is perhaps now time to recognise that the *evolution of complexity* is a central unifying feature in the relationship between the inanimate domain, organic life, and ultimately, conscious experience.

This shift in perspective can be likened to the shift in mind set generated by Copernicus' observations that the earth rotates around the sun rather than the other way around. It concerns, in the first place, recognising the relative nature of the observer's position and being prepared to look at events from new angles and beyond long-established assumptions. Superficially the difference may appear slight, but the consequences for our understanding of natural principles may be considerable.

Conventional views of evolution are largely based on the approach of *reconstruction* from laws of nature perceived and derived through the *deconstruction* of the physical forms of what now exists - particularly those provided by studies such as physics, chemistry and especially, thermodynamics. Many of these laws were formulated early in the development of modern science before recognition of evolution came into the picture. In my view, considering evolution in terms of complex self-organisation and hierarchical structure provides a much more comprehensive basis for understanding its expressions. This is particularly the case when we come to consider the more subtle domains of conscious cognition. Cognition, or mental functioning, is a widely expressed feature of sentient life which exerts a major influence over the directions taken by organic development, mediated through the normal processes of natural selection. Later we find the emergence of conceptual thought as a powerful influence in the domain of human culture, with consequences for all the higher domains of evolution.

This creates a new way of considering questions regarding directions evolution may be taking now - not so much in terms of biology but rather, of hierarchy, and this is a major element in the considerations of the remaining chapters.

The habit ever since Darwin, has been to view evolutionary time scales in terms of change over generations. A point I want to make here is that with the knowledge we now have, we can say that while genetic change is integral and essential, it represents just one domain of evolutionary self-organisation, whose time scale of perceptible change can usually be gauged over thousands or millions of years to produce significant alterations to (complex) species. If we think of stellar evolution on the other hand, its time span of gestation is mostly gauged in hundreds of millions or billions of years. Considered in terms of hierarchy, the principal medium of evolutionary development of human mentality is not to be found at the level of biology or 'organic' ecology, instead, we need to look at what one might call the 'ecology of culture' which is in effect a domain of higher influence still whose time frame for the creation of perceptible change is very much shorter - historically speaking, thousands of years, then hundreds, now perhaps even decades. Here we are talking about evolution within the collective field of human experience as distinct from that of biological evolution per se - although there are continuing consequences at that level - and as something in which we are fully active participants. In other words, the time scale of evolution should not be thought of solely in terms of millions of years past or the unforeseeable future, but as happening right now, at the scale of experience within our domain of human culture, and this shouldn't be regarded as trivial or superficial in our understanding of evolution. We are conscious participants in evolution, now, at the highest point in the hierarchy of domains to date, carving out the future of human nature, in the present, and consequently, for the rest of life and the condition of the planet.

Summary

This chapter considered several factors concerned in the idea of natural selection. Most of these, such as mutation, variation, adaptation, and ecology are fairly well-known and so I don't dwell on them too much.

What I do consider further is the idea of competition and its relationship to co-operation, and the representation of these qualities in nature generally. This is because these concepts greatly contribute to the way in which we interpret events in nature at all levels, animate and inanimate. There can be no doubt that competition is a major factor in development and refinement, but this factor is so distinctive that it has tended to dominate discussion to the exclusion of recognition that co-operation is an essential partner in developments. The idea of co-operation represented here is not that of deliberate action, but rather of innate integration, which is expressed from the most basic, sub-atomic, level of physical organisation to all subsequent levels. At any level of observation, it can be said that 'competition' exists between coherent entities which are the result of prior organisation, i.e., 'co-operation'. More generally, this dynamic balance between co-operation and competition, at all levels and scales of evolution, generates the context in which the emergence of all new form takes place, from the realm of sub-atomic particles to that of conscious experience.

Following up on that discussion I consider the transition from inanimate molecular chemistry to the emergence of LUCA, a single cell which has been deduced to have been the common ancestor of all known subsequent life on earth to the present day. This transition is a subject of much inquiry and there is a particular line of inquiry which I touch on concerning a form of partial isolation that takes place spontaneously and widely in inanimate nature. Partial isolation of various different kinds may well represent mechanisms that contributed to bridging the gap between inanimate molecular chemistry and cellular, organic, life. Phospholipids are of particular interest in this regard.

Chapter 5

Consciousness

Outline

Consciousness and intelligence are both important to consider because they clearly bear a great influence on the directions evolution takes. They are also of course, the basis of our ability to observe, and interpret that observation. In this chapter and the next I consider the relationship between these two qualities. There is a general impression currently that consciousness arises as a result of intelligence. In my view the evidence suggests the opposite to be true.

It could be reasonably said that consciousness is one of the most difficult concepts to provide definition of. One reason for this is simply that however one attempts to describe, define, or explain it, it represents a most inclusive basis for all experience, and as such has no opposite or comparable other than unconsciousness or non-consciousness, with which to give it context - its context is all experience. For that reason, I interpret it as simply, 'experience', in the sense of direct cognizance of 'existence', as the essential basis of all the forms of cognition in which it may be found expressed. In those terms it may be said that all functions of cognition throughout animate life forms evolve in complexity, with consciousness i.e., direct experience, as the initial and ongoing essence of that development.

Accordingly, intelligence as generally conceived of - as the faculty of reasoning - may be seen to be an evolved and evolving emergence of cognitive complexity arising as a result of the presence of consciousness widely expressed throughout animate life, rather than the other way about, which is a more conventional view at this time.

It can be further observed that consciousness, as the essential ingredient of 'experience', is at the root of all dimensions of 'experience' such as pain and emotions, as well as logic and rationality, and these dimensions should be taken into account in any holistic understanding of the meaning of 'intelligence'.

This chapter is one of the most challenging in this work for both the reader and me. This is partly because the ideas I want to offer are quite unconventional by Western standards, and partly because finding the words to express them is not easy. Nevertheless, this is an extremely interesting area to think about, and an important step towards topics I want to approach in Part 2, concerning evolution in the domain of human culture.

Clearly, consciousness and intelligence are significant topics in considerations of evolution, first of all, because they represent major features of life's development - as significant perhaps as the transition from inanimate nature to life - and are expressed at the current highest levels of the evolutionary hierarchy; and secondly these two qualities, or features of life are plainly intimately related, but the nature of that relationship is by no means clear, so it seems entirely appropriate to inquire into it.

Now that computers and artificial intelligence occupy such a significant part of our lives directly and indirectly, discussion of the meaning of consciousness and its relationship with intelligence has become much more common than it possibly ever has before. Previously it would have been of concern perhaps in more rarefied areas of philosophic consideration, now it's hard to read a newspaper or magazine without coming across some article related to this topic, albeit with the focus on questions such as whether artificial intelligence might lead to the emergence of consciousness within machines that possess analytical potential far beyond our own... There is a habit here of considering future possibilities based on existing assumptions, but without paying much attention to the basis of those assumptions. In this case an important one is that consciousness is emergent from intelligence. I look at these concepts, consciousness, and intelligence, what we mean by them and how they relate to each other.

I'll begin by observing that if the term 'intelligence' relates to the way information derived from experience is used, then clearly, the acquisition of raw information through 'experience' precedes that. In my view, the term 'consciousness' relates to this factor of 'data acquisition' - experience - rather than its interpretation and the uses to which it is put. The ability to experience - consciousness - is important to consider in terms of both its contributions to developments in the past and the directions these may take in the future. For those reasons, in one way or another, it becomes a prominent thread of inquiry throughout the rest of this work, underlying and more significant really, than intelligence per se.

First, what is meant by the concept 'consciousness'? Considering how central consciousness is to everything we know, it is a remarkably slippery term to pin a meaning to and can be found characterised or defined in a variety of distinctly different ways. Here I'll explore and outline as best I can in words, what it represents to me, in the context of evolution, and in the course of doing that I'll be presenting several observations and arguments to support my views. Once again, I'll ask the reader just to consider and weigh them up...

The current convention of science towards inquiry into the nature of consciousness is that it is an emergent result of higher complexity brain organisation, and to examine this organ to try to find any area/s or kind of behaviour that may pin consciousness down in terms of brain function. This appears to make good sense, but it makes the assumption that this feature we call 'consciousness', is not represented in cognitive faculties of animate life forms prior to later stages of the brain's evolution. The view I'll offer is quite different.

While there are plenty theories concerning consciousness, despite much experimental and philosophical inquiry over centuries, no truly testable, empirical, descriptions, of consciousness have been created, which means that all such ideas remain speculative. The approach I employ here is equally speculative but takes an evolutionary perspective. To begin with, when I think about consciousness, I naturally consider my own experience of what this means, and the sense of my own existence is a dominating impression. For that reason, I tend to associate consciousness with self-awareness. This is not unusual, there is something of a history of considering consciousness to be synonymous with self-awareness, but looking a little more carefully, it seems to me there is another way of viewing this that suggests self-awareness is a secondary development of conscious experience.

Most animate organisms express behaviour that indicates that, through senses that are mostly not very different from our own, they experience and respond to their environment, in other words, they are 'aware'. Clearly there is a great difference in the nature of the mental faculties of a human being compared to a fly or a spider, but each is also clearly 'aware' of its surroundings.

Normal everyday observation tells us this. To me, this quality, 'awareness', represents the essential meaning of consciousness. It seems to me that self-awareness is an additional feature of awareness, by virtue of which we experience our own existence, because of a particular faculty of perception that we possess but most other organisms don't appear to. For this reason, I would distinguish consciousness from self-awareness, and consider that consciousness is a more underlying feature of animate mentality, from which self-awareness is an emergent quality.

As in my own case, the usual starting point for consideration of consciousness is that of the form of human experience, which is reasonable since it is the most immediate and familiar expression of it for us to observe. Observation, logic, and reason appear to indicate that the brain is the essential organ that generates the experience of consciousness. Accordingly, we deduce that organisms that don't possess a brain, or have a much simpler one than us can't and don't express consciousness. However, there is also a historical background to this view, in which human beings were considered to have been created as fundamentally different from the rest of living nature, evolution was unknown and the brain, human or otherwise did not figure. Prior to the insights introduced by evolutionary theory, consciousness, with its strong association with self-awareness, was considered to be a uniquely human characteristic, along with intelligence. Despite our growing understanding of evolutionary processes and the evidence of behaviour in the animate domains, considerable traces of this view persist in our thinking and assumptions. But although the form of human experience appears highly complex compared to other organisms, we now understand that this form has evolved through many, many, degrees, and rests on a deep history embedded in the wider cognitive developments of nature. This view has become amended to the position that although intelligence is recognised to have evolved over long periods of time as an integral feature of nature, we tend to still regard consciousness to be a relatively recent emergent result of complex brain function.

Going back to basics, and looking beyond the idea of consciousness as unique to 'man', I offer the broad definition, *experience*. By that I don't mean memory of past experience, which is a way in which the term 'experience' is often used. Here I am referring not so much to specific incidents, occasions, or memory, rather, more directly, to the continuous *act of experience* that permeates and is the basis of, the sense of 'existence' (and consequently all those particular experiences archived by memory). By this I don't just mean our sense of existing as a discrete entity, or 'self', but more immediately and preceding even that, our sense of 'existence' of everything that we 'experience' through our senses of touch, sight, sound, smell, taste, and in more subtle areas, such as our very thoughts, which we also experience as 'existing' although they are entirely detached from our sphere of sensory experience per se. This sense of 'existence' is so basic, immersive and as-it-were integral to our being that mostly we don't notice it...! It seems to me that this feature, *experience*, doesn't only belong to human beings but can also be reasonably assumed to belong to any of the wide variety of life forms that display some form of cognition - sometimes referred to as 'sentience' - with which we share common ancestry we have evidence of going back around 580 million years. At the same time, it has long been apparent that human awareness differs from that of many other creatures in a sense I touched on, which is that we are aware of our own personal existence, i.e., self-aware, in a way that most others don't seem to be (I'm not sure how accurate that impression is, the more we find out the clearer it becomes that many other species are self-aware to at least some degree). There can be no doubt that self-awareness contributes considerably to our view of existence and the ways in which we interact with our environment, and I'll be discussing this further in a later chapter, but even so, in the terms I am outlining, self-awareness is a secondary, emergent, expression of consciousness, not consciousness itself.

Sometimes we become more overtly aware of being aware. A comparison might be the way we experience light. Mostly our attention is captured by the objects it illuminates, with our appreciation of light itself somewhat secondary to that. As we experience the objects it illuminates, we may notice the presence of light as an active source of that illumination by virtue of its variations of intensity, or shadow etc. If we are directing the light source, we will be aware of that also to some degree. Another analogy I can offer is that of a fish in water. Water is the fundamental medium of a fish's physical existence but is so all encompassing that it's likely that the fish entirely takes it for granted, without any direct awareness of it as such other than through its more overt perturbations. Nevertheless, the fish is evolved in form and function by virtue of its immersion in it. In this way, the fish can at the same time represent and reflect that medium and be largely unaware of its relationship with it because of that total intimacy. Similarly, although it is ever present, our 'experience of existence', of ourselves and everything else is so integral, immersive and immediate that normally we don't pay it much attention or recognise its all-inclusiveness as the medium in which we 'swim' - that which we call 'consciousness' - unless it's brought directly to our attention. Even then, in attempting to characterize 'it' the tendency, compulsion even, is to associate it with the particulars of the occasion, because from an evolutionary perspective, that's its purpose, to focus on the contents of experience rather than the 'act' of experience itself.*[1]

I've begun this piece by attempting to distinguish consciousness as a central principle of 'experience' from the range of expressions that experience may take, using the analogy of light. As I proceed to discuss the *forms* in which consciousness can be found expressed therefore, I'm referring to spheres of perception that have emerged throughout animate life. So far in these observations the starting point has been *human* awareness because this is the form of expression of consciousness with which we are most intimately familiar, but clearly the expression of our awareness is a complex evolved form which contains and represents many prior stages of progressive development that continue to function together simultaneously and can be seen expressed widely and to varying degrees throughout nature. It should be recognised though that understanding of the expression of consciousness, scientifically based or otherwise, is still rudimentary, and as ever, is based on the observation of certain indicators, which may not always be interpreted correctly. We can't say anything here with certainty, merely use observation and deduction to try to put things into a reasonable order...

Self-awareness is generally considered to be an expression of consciousness found only at the higher end of observable intelligent behaviour. For that reason, it is often taken as a gauge of the intelligence status of particular creatures or species. The methods we have of exploring self-awareness are not particularly good however and don't really reveal much about its breadth of representation in animate life. The reason for it being taken as a prime indicator of consciousness is down to the fact that our observations lead us to associate it with the presence of a high level of abstract cognitive functioning, such as we humans have, and simple experiments that demonstrate its unambiguous expression in some other highly evolved species such as apes, elephants, whales, and some species of birds. In other words, we use our own, human, status as the principal reference point for evaluating the expression of consciousness and intelligence elsewhere on the spectrum of animate behaviour. This is natural and reasonable - to base observation around what we do have familiarity of - but the deductions we make from this position alone can be very limiting, and potentially misleading. This is partly because of the weight of historical assumptions, which I've touched on and will discuss further, partly through a failure to recognise the basic principle that complex forms always develop from simpler precursors which are generally much more widespread and may not be so immediately obvious. This applies to the dimensions conscious human experience takes just as much as it does to anything else.

Viewed in terms of evolutionary progression, self-awareness represents a secondary emergence from a broader domain of what we could call 'abstract' awareness, and may not be entirely unique to higher species, but may exist to some degree wherever abstract awareness is present. As we interpret the world at large in abstract form, so also do we experience and interpret ourselves as 'objects' of our own perception. Abstract awareness is a sphere of cognitive function in which mental representations of experience can be manipulated to some degree, in pursuit of goals of survival and propagation. This ability appears to be present widely to some degree throughout animate life forms - and by that I really refer to domains of life in which cognition at some level is evidenced by the presence of a nervous system. The human species may express it to a much greater degree than can be found elsewhere, but by no means uniquely. One feature that *does* appear to distinguish the human version of abstract awareness is the faculty of *conceptual thought*. But conceptual thought perhaps shouldn't be thought of as a 'higher level of awareness', so much as a 'tool' or range of tools exercised within the jurisdiction of abstract thought. Conceptual thought is of course deeply linked with language and has been highly developed as a human attribute over time, but other complex forms of abstract manipulation that we aren't equipped to recognise easily may well exist for some other species. Also, if self-awareness is a result of abstract cognition, its nature is intimately tied into the forms of that cognition. In other words, the form of self-awareness is shaped by the filters of cognition through which we perceive ourselves.

What can be reasonably surmised is that faculties of abstract cognition, without the additional attribute of conceptual thought, and regardless of the presence or absence of self-aware-ness, are already represented across a broad spectrum of animate life-forms. This is evidenced particularly through the ability to learn from experience and refine behaviour. To my mind, a good example of this is represented in the way birds control their flight. Certainly, a large part of the control involved is inherited as innate instinct, but a young bird still has much to learn to become expert at manoeuvring and surviving in the many different and varying situations it will encounter, from the practicalities of flying to dealing with predators and prey. Recent experiments have shown that many animals, from ants to whales possess a range of essentially similar complex cognitive skills involving learning in dealing with their environment. As human beings, this faculty remains an integral aspect of cognition that we use all the time. This level of abstraction provides the ability to evaluate, predict and act directly on immediate experience without the requirement of conceptual thought or analysis. There is clearly cognitive ability involved in these tasks that appears to depend on the manipulation of abstract information - thought - but at a more intimate and immediate response to sensory input than that involved in planning or calculation that we as modern humans know it. (Although in some societies, this level of abstract thought is much more directly exercised than in others, such as our Western culture, which is more immersed in conceptual thought. For this reason, it seems clear that there are cultural dimensions to the degree to which each is exercised). This level of abstraction then appears to extend further in many higher species, to an ability for short term planning, still without the aid of conceptual thought. One might think of the behaviour of dogs, cats, apes, dolphins, crows, and other birds, for easily accessible examples. It may be helpful then to consider this faculty of abstraction as a distinct feature of a wide range of animal behaviour that expresses it, all of which can be considered abstract thought, but which utilises this faculty to varying degrees of complexity. This expression ranges through those I've touched on here, from immediate response calculation at one end, to conceptual thought at the other. It should be clear though that, as ever, the more complex levels still depend on the retention and support of the prior levels. Conceptual thought, for example couldn't work alone without the presence of the lower layers of immediate response evaluation and non-conceptual calculation. These occur not only simultaneously but in an integrated way. This, again, reflects the intimate and hierarchical nature of their relationship.*2 It's worth observing also that this general faculty of abstract awareness most probably had its origins early in the common ancestry of all *animate* life forms unless it emerged independently in different contexts. This is possible but unlikely.

Seen from this perspective, the central principle of 'experience', i.e., consciousness, is common to these different expressions of awareness, while cognitive functioning provides the actual form experience takes. Returning to the original question, what is consciousness? In these terms consciousness may be viewed as this central, inclusive, phenomenon - 'experience' itself - manifest first through the direct experience of the senses, next in the mental experience of abstract association in all its degrees of expression - i.e., non-conceptual thought, then by conceptual thought as a further, emergent, factor. Any form of self-awareness will reflect the degree and nature of abstract awareness but doesn't depend on the presence of the ability of conceptual thought and can certainly be recognised as being present in creatures which don't possess that faculty. Conceptual thought, however, is unlikely to be found in the absence of self-awareness, as it indicates an already advanced level of abstract perception.

None of these considerations help to 'explain' or define what consciousness is, any more than light is explained purely by describing how it illuminates objects, but they may help to approach an understanding by outlining the relationship of different aspects of its expression in the human attention and more widely in sentient nature.

Dimensions of Experience

In these days we have created machines that can register and respond to stimuli in many areas we associate with the senses, particularly touch, sight, and sound, and to carry out complex calculations. But do we consider them to be 'conscious'? We don't, any more than we would consider a calculator to be conscious; however complex its design and calculating skill, its functional ability remains essentially mechanical. So, what is the difference and is it possible that computers could 'become conscious' eventually once their ability to process information had reached some particular level of complexity? I believe this question arises mainly from a tendency to conflate the ideas of 'consciousness', 'intelligence' and 'self-awareness'. According to the view I am presenting, complex intelligent behaviour and self-awareness are later developments from abstract awareness. For any machine to become 'self-aware', on this premise it would have to be aware - i.e., conscious - in the first place.

However, many people involved in the sciences, particularly computer design and IT, suppose that consciousness is a later, emergent, feature of 'intelligence' where intelligence is generally understood in terms of the ability to manipulate and analyse information on a broad spectrum and that as the development of artificial 'intelligence' progresses, consciousness may well emerge as a result. From the perspective I am presenting here, this view of 'intelligence' seems to me to be superficial and provides a very limited basis for the explanation of certain other dimensions of experience which appear to be widely expressed in nature but clearly aren't a result of intelligence in that sense. One obvious example of this is 'pain', which appears to be an experience available to a wide range of animate creatures regardless of their ability to 'reason'. Now, many of those who approach this question from the conventional 'intelligence' angle - and there are many - suggest that pain doesn't exist as we know it for creatures of lower evolved intelligence, although we may mistakenly infer it from observed reflex responses and the presence of a nervous system. To me this appears to be based on the inverted reasoning that the experience of pain involves emotional content, which requires the presence of consciousness, and consciousness is an emergent result of intelligence, therefore no creature without the faculty of 'intelligence' can experience pain, regardless of how they may react to injury. From this point of view, response behaviour is regarded as being entirely 'instinctive' reflex, devoid of any 'experience' content, which emerges later in the phenomenon we call consciousness...*3 My understanding is that this line of reasoning derives from the old Western theological view of a fundamental distinction between the human condition and the rest of nature, rather than the actual evidence of observation...

To my mind it seems clear that pain plays an extremely important protective role that confers significant evolutionary advantage across a wide range of animate life and works because of its immediate potency as *experience*. The experience of pain clearly precedes the emergence of analytical intelligence and supports the impression that the factor of experience precedes the emergence of complex intelligence - and not only the intelligence represented by conceptual and analytical thought as we represent it in the human domain. I would suggest that this, the factor of experience, also applies to other forms of 'intelligent' behaviour such as for example the highly organised collective behaviour of ants and bees and other such creatures which express experience and complex organisation, but without recourse apparently, to conceptual thought. Individually their behaviour may not appear to be particularly 'intelligent' but close observation of their behaviour, as I touched on earlier, has revealed surprising cognitive skills including, for example, the articulate identification and use of use of simple 'tools' by individuals in quite complex activities. Both ants and bees have been observed using tools 'cleverly' and on close inspection, to display unexpectedly complex cognitive and social skills. 'Intelligence' in these cases is often most apparent in the context of collective behaviour, but it is also apparent in the behaviour of individuals as demanded by circumstance. Again, to my mind, this is easier to understand if one interprets it as intelligent behaviour arising as a natural result of the expression of conscious cognition, rather than the other way round.

Other areas of experience to which the observation regarding pain also applies, are those of all forms of emotion. These bear a similarly important influence on the development of evolutionary form but are often regarded as less relevant than the development of logical intelligence. If, however, we consider consciousness to represent the essential principle underlying *all* dimensions of experience, including pain and emotions of all kinds as well as sensory experience, then plainly consciousness is widely expressed in living nature much more broadly and at levels much more fundamental than that of 'reason'. The fact that machines can carry out analytical processes that in many cases equal or surpass those of the human mind illustrates that the mental faculty of logic that we tend to hold in particular esteem simply reflects the objective quality of causality and is, itself, essentially mechanical, however sophisticated its operations may become, rather than in any way indicating the presence of consciousness as a unique, higher, faculty. In these terms, the refined development of this faculty of logic in our species is that of a useful, practical, tool which is in itself devoid of any 'superior' quality of conscious content. In our case, it has evolved to occupy a prominent position through the presence of many other qualities and features of experience, along with which it has evolved over great periods of time, and which are entirely absent from machines.

Analytical intelligence, in these terms, can be seen therefore to be an emergent expression of consciousness, but not alone. I do think that intelligence and consciousness are inseparably related in the same way that a branch is inseparably linked to the trunk of a tree, but I believe that it is a major mistake to identify the concept of 'intelligence' only with the faculty of analytical reasoning, which is historically how it has tended to be identified within Western convention.

Like many people I'm not over fond of spiders but I am impressed by their articulate skills in web-weaving and by their acute awareness of their surroundings, and I'll borrow that impression to illustrate my perception that consciousness appears to be manifest in many creatures sharing common ancestry with us going back a very long way... It seems likely to me that for any creature the possession of a nervous system, however simple, indicates the ability to experience, i.e., consciousness, within the terms of its cognitive capabilities, however restricted or expansive those may be. This ancestry is very ancient, still, an implication arises that at some point there was either a shift from non-conscious biochemistry to the domain of experience, *or* that some underlying quality of 'experience' existed already at some level in organic life before its first recognisable expressions in overtly sentient life forms.

A chief difficulty in relating to this broad view of consciousness is, again, the tendency to identify it with the human form of its expression, which, naturally, is our starting or reference point for inquiry into its nature. As we retrace its emergence back to its evolutionary roots, we have to try to distinguish consciousness, as a natural principle, from the variety of cognitive forms that express it. This could be likened to unravelling a richly patterned fabric while recognising the unity of the thread with which it is woven.

The Evolution of Cognition

Before proceeding further, at this point I should try to outline a little more, what is meant by cognition. Cognition refers to the range and instrumentation of the varieties of mental experience expressed throughout all complex, animate, life forms. Extending an earlier analogy, if consciousness is compared to light, 'cognition' could be compared to the instrumentation of that illumination, from torch bulb say, to chandeliers, to lighting rigs, to fibre optic cables, and the 'structural frameworks' of the regions illuminated, though not the actual content illuminated. This relates to all the vast variety of sensory experience to be found throughout animate nature, the *interpretation* of such experience, and *response* to it. (An analogy could also be made here with a computer system, where there is hardware for processing information, accompanied by 'operating system software', which together provide the central systems of cognition, while sensory links with the outside world provide the raw data, or content, to be handled.) These dimensions of cognition constitute what we refer to as 'mind'. Within these terms, 'interpretation' is essentially what we refer to as 'perception' and broadly speaking is implicit wherever the presence of a nervous system is identifiable, however simple.*4 This applies also to 'response', as the mental basis of all forms of 'action' from the most elementary to the most complex. This includes every animal from the earliest 'primordial worm' to the human being and includes thought of all forms as a subtle form of action. The nervous system is the principal medium of perception and action and is considered in (a little) more detail in Chapter 10.

As with all evolutionary phenomena, the development of cognition has been far from linear. While the instrumentation of cognition rests on foundations of physical principles, such as the action of light, sound, electrical charge, gravity, biochemistry etc, no species or the forms taken by their physiology and cognition, evolve in isolation from their ecological context. This relationship is what is indicated by the term 'holistic' and is all-encompassing. Even in considering non-sentient life forms, in terms of biochemistry alone, this principle can be applied all the way back to the earliest emergence of the precursors of the first form of life as we know it. In evolution context is everything. It is entirely dynamic and vastly complex, far beyond the scope of analysis down to simple components. This doesn't mean that understanding of the function of those components is irrelevant, but what it does mean is that no phenomenon produced by evolution has an existence that is independent of context. When we consider cognition it's important to bear in mind that here we are considering the evolution of the instrumentation of perception in conjunction with the evolution of physical form, or physiology. The two go inseparably hand in hand, and I'll explore this a little now.

The following argument may be a little unconventional...

While seeking to avoid viewing these matters too much in human terms, if we recognise that the structure of our own psyche reflects its evolutionary history, it's worth starting off inquiry into cognition by observing aspects of our own thought processes and behaviour, which represent perhaps the most immediately accessible area to explore this topic.

An example I find useful is this; we can see that in everyday life activities, such as learning to drive a car, to play a musical instrument or use a tool, we learn through cumulative experience, which normally requires a combination of attention (directed awareness), action and repetition. Thereby, memory and habits are formed which no longer require full attention - some attention is usually still required but this may often be quite minimal, or residual - you might say, supervisory. So, a layer of behaviour is consciously initiated at the level of response to direct experience, which then becomes established at a level requiring minimal consciousness - effectively acquired reflex. This is a well-known and familiar principle of everyday acquaintance, broadly represented in terms such as 'learning', and 'muscle memory'. If we now consider genetically established innate reflexes, or instincts, the general approach of evolution theory is to default to the notion that all such behaviour is ultimately the result of the cumulative trial and error selection of genetic mutations, and that acquired skills can't be transmitted genetically from one generation to the next. But there may be a more subtle, and relevant linkage between acquired skills and inheritance that I'll explore here a little.

While it has long been clear that specific attributes or

skills acquired in the lifetime of any individual are not passed on through genetic inheritance, an important factor in the establishment of behaviour within a species is its spreading through social sharing. This takes place within all higher evolved species to some degree, particularly through parental 'instruction', but also in wider interaction within their community; extended family, peers etc.*5 Therefore, although behaviour acquired by invention or learning is not genetically recorded, the factor of collective learning reinforces the long-term influence of acquired behaviours within a species over many generations, with associated effects on physiological developments and characteristic that distinguish that species.

In effect this means that in domains governed by cognition - and social sharing particularly - the inheritance of behaviour and physiology is very much more complex and multi-layered than can be described and explained by mutation and molecular genetics alone - always a function of the relationship between the organism and its complete environment, mental and physical. It seems to me that this could link to the development of inherited instinctive behaviour in a way analogous to that referred to above in the example of acquired reflex. This would take place on a cumulative, microcosmic scale over many generations, via the intimate relationship of cognition to physiology, rather than as the genetic transfer of learned behaviour directly from one generation to the next.

In this regard therefore, it seems to me that the evolution of cognitive skills, instinctive motor skills and overall physiology have all gone hand in hand, with conscious experience, in the moment, being a dominant, directing, influence in these relationships, and the inheritance of characteristics isn't confined to physical ones. Clearly, cognitive skills are very important in the scheme of things and, intimately interwoven with physiological features, are selected for accordingly.

The tendency to perceive and respond in ways that prove selectively advantageous in certain kinds of situations, reinforced through social sharing, would establish its representation at the genetic level, along with any associated physiology, complemented and refined over generations by mutation and sexual variation. Another way of putting this is that cognitive skills evolve in close linkage with, and reinforce the value of, 'instinctive' behaviour. The effects of variation introduced through mutation and at the higher order of sexual exchange, are amplified by this context; otherwise relatively insignificant variations may become established on account of the use accessible to articulate cognitive faculties (clever skills). Consider goats' hooves; a principal skill of mountain goats concerns their agility and deftness in balance and control in dealing virtually instantly with complex terrain, the nature of which is a feature of local topography. Even very minor variations in the texture of their hooves may make a significant difference to their suitability for particular kinds of surface. Without the highly developed cognitive context, such variations would be of much less importance. The cognitive context bears a high degree of relevance to the consequences of mutation and the establishment of instinctive behaviour, and this context extends into the sphere of collective, social, behaviour.

When an animal such as a cow or elephant, giraffe or zebra is born, it does so with established instinctive motor skills that enable it to use its legs very quickly to walk, but it still must further develop those skills through practice. This is similar to the example I offered earlier of fledglings learning to fly, they are born with basic motor skills but have to develop fine skills through practice. All these skills originally developed in circumstances where predation would select out the slow and less agile very efficiently. In practice then, survival skill would be based on a combination of inherited/genetically established motor instinct and cognitive ability - those that learnt quickly would have the advantage. Again, this puts evolutionary pressure on cognitive ability, which becomes selected for in a priority sense. The main point here is that if inherited innate physiological skills, such as motor reflex actions, were to depend on random selection *alone*, this would take very much longer to produce results - if ever, and never to the same degree of complexity.

Hierarchy is important to recognise here. Mutation constantly provides new minute variation essential to constant evolutionary 'experimentation', but its position in the train of hierarchy is necessarily low, at the level of genetic molecular chemistry. Cognitive context is also genetically prescribed but is in a much greater position of influence hierarchically and its variations are contributed to much more through the higher order medium of sexual exchange for example, than through mutation.

Weaver birds perhaps offer an example of another important aspect of this relationship between cognitive skill and innate motor instinct, which is active sexual interaction/co-operation. These birds, like all birds, build their nests with a strong inherited instinctive component, with a variety of complex overall patterns represented across the range of their species. But this is combined with a constant element of unique individual input that involves learning through practice. As with learning to fly well, which is necessary for all birds, learning to weave well requires much practice. This is a point at which the engagement of conscious attention takes place, with consequences for the direction of subsequent developments, both for the reproductive success of the particular birds involved, and the cognitive heritage of their genetic line. This is the microcosmic level at which change takes place; change is microscopically gradual. It is the result of the individual conscious actions of the birds in question, in the present, but entirely out-with the range of any conscious pattern of design on their part. Clearly the major patterns didn't occur overnight, but are the result of many, many generations of naturally selected 'most effective' design, nevertheless, they are continuously being refined and experimented with by each individual bird. Repeating an analogy I've used elsewhere; at the level of the microcosm each effective engagement adds a 'pixel' to the macroscopic 'image' of the species in question. In a very real sense, each bird, through its conscious participation, is not only weaving its own nest but contributes also to the future shape of its family/species. Again, the usual convention of evolution theory is that this results from the cumulative selective advantage provided by purely random variations, with no input from acquired experience. But there is clearly much more to it than that.

To expand a little on the context, first, in this case the nests are built by male birds, and as is most often the case in such circumstances, this has to do with advertising their reproductive suitability to the females of the species. While this is sometimes regarded (by human observers) as a measure of general physical health, equally significant qualities being advertised here are effectively, *cognitive skills* reflected in the quality of the design, and no doubt, detail, to be accepted or rejected by the females. Clearly there are two sides to this, because, on the one hand is the active cognitive functioning of the male, expressed in the design and construction, on the other is the perception on the part of the female, of the desirability of the design and of the cognitive skill of the male. Both cognitive dimensions are involved and reflected in the selective outcome. This represents something of the importance of the contribution made to the evolution of cognitive skills through sexual co-operation. Two quite distinct areas of cognitive ability participate and evolve together in a complementary way. This principle is likely to have been of major importance in the evolution of higher life forms and further illustrates that there is selective linkage between active cognitive skills and the ongoing formation of instinctive behaviour. You could say that the cognitive sphere becomes a significant contributor to the evolutionary environment for the more 'mechanical' biophysical sphere. Overall, both the innate physiology and the cognitive skills benefit, and are mutually sustained through selection.

It can be seen then that although cognition may not produce directly inheritable effects on physiology or instinctive motor reflexes, its contribution to their shaping is considerable. It seems reasonable to expect it to have been an important element from the very earliest stages of its presence in animate life forms. The nature of this relationship parallels that between conscious learning and 'muscle memory', or conditioned reflex motor control, which I mentioned at the beginning of this piece. The cutting edge in each case is conscious cognition, directing action in present behaviour that results in; on the one hand, learned reflex confined to the lifetime of the individual, and on the other, *instinctive* reflex established and refined over generations by natural selection. Because of our everyday acquaintance with the first example - the relationship through learning - the comparison may be useful in exploring the evolutionary linkage between cognition and physiology, above and beyond that provided by genetic mutation and sexual variation.*6

If we consider all these phenomena in terms of evolutionary hierarchy, with cognitive skill highly positioned in the pattern of developments, it is clear that the ability to *share* learned experience plays a significant role that is relevant to the areas both of the social establishment of newly acquired skills, and of ongoing physiological evolution of the species. Where the ability to share learning is well developed, the sphere of collective behaviour then becomes a dominant influence in the hierarchy. This is certainly the case for human society, and very probably also for creatures such as whales, apes, and similarly highly evolved animals. What then, of ants and other creatures that we see to be highly collective and social in their behaviour, but which don't appear to express much in the way of sharing new experience? Recent experiments with ants and bees have shown that both of these have the ability to learn through the observation of others and employ individual ingenuity when called on by circumstance, in overcoming obstacles to carrying out their duties, which indicates that although their behaviour *appears* largely bound by strict instinctively established protocols, nevertheless some degree of mental skill and flexibility is present, and probably essential, in their individual behaviour, not just the mechanical following of programmed instruction. One should bear in mind, that faculties involved in ingenuity are also inherited, so it's not as if they exist in any way external to the selection process.

It seems to me that in the terms I've been portraying, in such cases although individual cognitive input has been demonstrated experimentally to be present, and probably contributory to some degree, the collective relationship appears to be much less influenced by that individual cognitive input than it appears to be in the likes of human society. An analogy that springs to mind here is that of soldiers in a highly organised army; individual soldiers are constrained by strict rules of collective behaviour, but nevertheless, even in the strictest of such regimes, the integrity of collective behaviour must rest to some degree on individual ability to respond to and deal effectively with circumstances of the moment as they arise, even though the skill set and scope of decision making of the individual may be very limited. In the context of human behaviour, that would translate to, for example, skill in interpreting and evaluating situations appropriately as they arise, ability to communicate clearly, to act decisively and effectively. All these skills belong, in the first instance, at the level of individual behaviour, and are at the same time the crucial basis of successful collective organisation, however rigid or flexible that collective sphere may be. Without some degree of individual cognitive flexibility, it's hard to see how any complex social organisation based on cognition could come about. At this level, personally, I don't think that selection around mutation, or sexual exchange, could alone generate the very complex and sophisticated patterns of behaviour that are widely observable, although in effect this is the conventional view.*⁷ Regarding species such as ants, it's worth remembering that, mostly, a new generation cycle may take place anywhere between a few months and around five years, and this has been happening over a period of around 140 - 160 million years, as compared with homo sapiens for whom it's more like twenty years per generation over a period of around 200 thousand. Perhaps individual cognitive behaviour has become increasingly prescribed towards such dedicated patterns of collective behaviour over many generations, while at earlier stages it was much less defined and more flexible. (Not that I'm suggesting that in ancient times ants were individualistic in a developed intelligent way, just that the individual cognitive skills that they do possess may have been less constrained and defined by established routines, to act and contribute towards the creation of patterns of collective behaviour that have subsequently become their species signatures.)

The linkage between the evolution of physiology and cognition can be seen expressed throughout all animate life forms. This has consequences for all the other life forms that they interact with, above, below, and alongside them in the regions of ecosystem that they occupy. Clearly, wherever cognition exists, at any level and in any form, it is associated, directly or indirectly, with physiology of some kind, and selection pressures have been active in this relationship from the dawn of its emergence. Mostly this is fairly obvious when one thinks about it, and entirely un-controversial. My main observation here, as I've discussed, is that while genetic mutation provides the raw basis of variation, it is a considerable over-simplification to take that principle alone as being the entire answer to the questions of the evolution of complex life and the forms that have emerged. It would perhaps be more accurate to say that with the emergence of cognition, genetic 'mechanics' of reproduction and variation, which include mutation and sexual combination, have become partnered with cognition. In this partnership, viewed hierarchically, cognition is the dominant influence, genetics the 'sleeping' partner. This rela-tionship is at the root of the evolution of every feature of animate sensory apparatus and 'motor' action such as eyes, ears, vertebrae, fins, legs, arms, wings, etc. Viewed in these terms all these attributes have come into existence and been refined with the aid of the active (conscious) cognitive contributions of countless billions of individual sentient creatures over hundreds of millions of years.*8

Usually when we think of the social domain of behaviour and behaviour exchange, we tend to think of this as a particularly soft and transient layer of phenomena that doesn't contribute fundamentally to evolution at the level of genetic inheritance. However, when we consider evolution in terms of hierarchical organisation rather than focussing strictly on the genetic instrumentation of transmission, it becomes clear that this is a misleading impression. Bearing in mind the integration between successive levels, or layers, of cognition, the social communication of behaviour will help to sustain not only the behaviour itself but also the mental skills that gave rise to the behaviour in the first place. Simply, 'successful behaviour' locks the cognitive skill into a heritable status and social 'sharing' of behaviour reinforces this at a species level. Social behaviour will therefore always have an intimate linkage to the more deeply established level of genetic inheritance, while nevertheless depending to a great extent on transmission through learning and the sharing of experience. This places the social dimensions of experience at the front line of developments.

In the case of human culture, if we view it in the microcosm, the relationship between the individual and their environment is at the core of its 'substance'. At the same time, the capacity of the individual to shape culture directly is mostly very limited (not entirely unlike ants...). Nevertheless, the collective shape, locally and on greater scales is formed through the cumulative action of individuals over time. Expressed in terms of evolutionary processes, this represents the general principle of the relationship of the 'component' with the environment. In all domains, the 'components' are the essential ingredients whose innate qualities comprise the potential that is choreographed by the environment. Accordingly, pressures at the level of the environment determine the overall shaping of behaviour at the collective level - the nature of the choreography, which I'll discuss a bit further in the chapter on 'Cultural Evolution'.

To summarise these points; much animal behaviour appears to express a combination of genetically established factors, i.e., instinct and learning skill. Bird flight, which I touched on earlier, and many other kinds of animal behaviour plainly express this combination. Specific learned behaviour may not be passed on through genetic inheritance, but cognitive skills involved in learning certainly are. Active cognitive skills combine with and augment pre-established instinctive behaviours. Clearly, these faculties are also inherited, but developmentally, they occupy a sphere beyond that of instinct, which is traditionally considered fixed. Also, the social sharing of behaviour reinforces the conse-quences of this at the species level in terms of both mental faculties and physiology, and represents a significant *further* factor. At all stages, natural selection is the 'judge' of what works best, at both the individual and collective levels, and at this level of the hierarchy of life, whatever the precise mechanisms of inheritance may be, among significant qualities selected for are active, in the moment, cognitive skills and the ability to innovate constructive-ly.

Experience and Response

While considering consciousness to be the central feature of cognition, it can be observed also that there are two distinct aspects, or qualities, to cognitive functions, namely, *experience and response*. The relationship between these aspects can be seen to underlie the evolution and development of all cognitive form in animate nature, from the most elementary to the most complex. They can be recognised as principal participants in the dynamic structure of the human psyche as well as in the activity of, say, an earthworm, from the level of direct sensory experience immediately directing response, in the case of the worm, to the level of the complex dimensions of emotional experience and action, mental and physical, of higher evolved organisms. Although response ultimately involves some degree of physical action or process, at a subtle level it includes many forms of mental expression, whose 'shape' and form have evolved in accordance with all the circumstances and pressures of evolutionary conditions.

The range of emotions experienced by human beings is of course extensive and complex, but these emotions have evolved and been shaped from simpler roots that can be widely observed in animate life forms. Although these emotional dimensions can often be recognised from our interaction with many animals, from dogs and cats to elephants, mice to dolphins for example, putting them into words is another matter. However, my interest here isn't so much the detail of emotional experience, rather, the general principle of the evolution of emotional form, and its relationship to intelligence as we generally conceive of it in terms of rationality. One overt example of this perhaps is anger or aggression, which can be seen expressed widely in sentient nature. It clearly has survival value and is generally 'contained' within contexts of ecological balance, but my point here is that it is clearly a mental form arising from and 'composed' of consciousness. Show me a computer capable of anger... Such emergent creations of consciousness are important features of what we refer to as 'mind'. I chose aggression as an example because it represents an emotional state that we can recognise from our own internal experience. There are of course many other mental states that we experience that are also shared to some degree by other life forms, but can be more difficult to be certain about, or even define clearly from a human perspective, such as love, affection, and emotional distress. Anger and perhaps fear stand out as emotional states which are generally recognisable across a broad range of animal species. 'Mind' as a term really includes all dimensions of mental experience from the most extreme to the most subtle and from those of pain and the emotions to rational logic. This however suggests that what we refer to as 'consciousness' doesn't just mean 'experience' in fact, it also represents the essential source from which all forms within the domain of 'mind' emerge - not only the human mind, but mind as represented and expressed throughout all cognitive life forms. An inclusive term we commonly use in reference to this is 'being', as in 'human being' for example. Viewed in these terms consciousness is the primary ingredient of 'mind' while cognitive evolution is the medium of its expression and formation, highly 'plastic' and capable of generating a wide range of mental form.

I would emphasise a point here in case I haven't already made it clear. The conventional view of all mental form is that it is no more than a result of entirely physical processes of chemistry and biology and is therefore insubstantial, or illusory. In the view that I am presenting, the central theme of evolution is *self-organising complexity*, which transcends boundaries of description that we may characterise as physical or otherwise, because ultimately all such regions are transitional but contribute equally to developments. From this perspective, principal considerations are not bound by the definitions we make in any *particular* fields of observation, but concern relationships within the hierarchy of organisation. In this regard, mental form is high in the hierarchy of domains, exerting great influence, and, despite its entirely non-physical nature, or field of function, is therefore no less 'objectively' real than atoms, molecules, or DNA.*[9] Because of its distinct sphere of dominion, it is no more valid to consider mind only in terms of neural function or biochemistry than it is to think of living processes in terms only of molecular processes or chemistry. As ever, underlying domains are essential, integral, and thoroughly relevant, but their dynamics can't be used to explain entirely, the dynamics of higher domains. This is not to say that common features can't be recognised. The terms of description I use in discussing mind here should be regarded as, 'illustrative' rather than representing an attempt to be strictly 'factual'. For me, as I have said earlier, a great part of the interest is in finding features which are common to, and represented in all the domains we inhabit and share with nature, animate and inanimate.

If we take the central principle of 'mind' as being the experience of existence - i.e. - consciousness, and cognitive faculties as providing structure and 'dimensions' to the variety of forms of experience, their interpretation and response, the form of 'mind' may range immeasurably, from say, that of an insect to that of an elephant, a fish, or a human being, while the essential principles are the same. Remember though that from this perspective, the 'raw substance' of all mind, consciousness, has no shape, size, or dimensions, in itself, and in a sense is distinct from the cognitive forms to which it gives rise. An oft used comparison here could be to clay, as the material from which many different forms may be moulded while retaining its essential nature.*10

To help the reader to conceive of the presence of consciousness in creatures very different from humans, I'll extend an analogy of light that I employed earlier, this time as bright light, dim light, light in large spaces, light in small caverns. There has long been reference made to light as a metaphor for understanding, usually in the context of human awareness; here the image is intended to apply more broadly to consciousness as the essence of experience. The 'space' being illuminated is the cognitive 'space' of any creature, be it a human being or a spider, an elephant, or a snail. The level and form of 'experience' may vary immensely, but the 'light' providing illumination is the same. In these terms 'cognitive space' is 'mind', the 'light' is consciousness.

Cognitive Form

To recap briefly; in the context of human experience, the term cognition is mainly used in connection with thought processes. But this has a very wide range of relevance when one considers the fact that thought processes involve many different factors, such as memory, mental imagery, rationality, emotional input etc. as well as direct sensory input which provides the raw material for all of these and all the other established structures of perception that contribute to its interpretation. The term 'cognition' applies to all of the instrumentality of conscious functioning. Bearing in mind the evolutionary background of the human mind, for my purposes, 'cognition' also applies to this in-strumentality wherever its presence can be recognised in nature. In practice, at present this means, wherever a nervous system is seen to be present, however rudimentary, or complex.

The expression I am using here therefore, 'cognitive form', is in reference to the range of dimensions of 'experience' that appear to exist throughout animate life. I'm not concerned to attempt to explore the extent of this range in any detail, what I will try to do is draw some insights about it more generally against the background of human experience.

Returning to a point I touched on earlier in this chapter; sensors can be connected to a computer such that it can respond to light, sound, contact or proximity, to temperature, electrical charge, magnetic fields and so on. This we don't consider to represent or express consciousness; again, a computer can carry out very complex analytical operations, but we don't regard these as being conscious activities, merely as essentially 'mechanical' transactions. In many ways these electronic operations, based on the transfer of information in one form or another electrically, are very similar to processes present in nature, particularly in human beings, that we would regard as cognitive functions. So, what dis-tinguishes conscious experience from the purely mechanical, or physical, level of interaction/response? Another way of putting this question is; could similar life processes function effectively without the presence of consciousness? Certainly, none of these means of interacting with the environment depend on conscious-ness. In fact, entirely without suggesting any evidence of 'con-sciousness' many modern computers can carry out complex calcu-lations far beyond the scope of most human ability, which we would previously have regarded as intelligent operations and now often refer to as artificial intelligence. But there are other dimensions of experience involved in the expression of consciousness and intelligence in living organisms, which haven't been reproduced in computers, but which have played fundamental and integral roles in the evolution of those qualities. Considering these other factors may help to some degree to distinguish the mechanics of perception and information manipulation from consciousness itself. Examples, which I've touched on, are first, pain, then the emotional dimensions of experience. These aren't just curious incidentals of evolutionary events, these dimensions of conscious experience represent driving principles of evolution, exert great influence in the selection process among life forms and underlie the emergence of intelligence, however we may conceive of that. Without the contributions such aspects of experience have made to the evolution of cognition it is unlikely that evolutionary form would or could have developed beyond the stage of basic biochemistry, let alone emerge as functional intelligence.

In a sense then, the various instruments of sense experience - and even intelligence as represented in the faculty of logical reasoning - can be regarded as being largely 'mechanical' in function, however the fact that they have evolved at all is a result of contributions made by the wider dimensions of consciousness experience, which remain as their constant and essential background partners.

Motivation and Problem Solving

There is, perhaps not unreasonably, a tendency to regard rationality and emotional experience as being fundamentally different in nature, and also, for that matter, sensory experience. We don't see at first glance much in common between say, touch and anger, or love and logical skills. The one thing they plainly do have in common is that they all belong to the domain of conscious experience. All are illuminated, or enlivened, by consciousness. Considering this in terms of their origins and evolution some observations can be made about their relationship to each other.

Plainly, there are features common to all living creatures, from bacteria to human beings, that can be generally characterised as motivation and action. At a biological or organic level, motivation covers a range of survival impulses and motor functions embedded genetically that find their roots already represented in the deepest antiquity of organic nature at the cellular level, such as the search for sustenance, defence, and reproduction. The satisfaction of these fundamental needs involves action of one form or another. These basic characteristics of life - motivation and action - pre-date the emergence of complex, multicellular organisms, and cognition as we know it, but clearly continue to underlie and contribute to the dimensions of form taken in their development. Observing at the level of simplest organisms, this may be regarded in terms of relatively straightforward biochemistry - the direct absorption from the environment of the nutrients required for metabolism, but the growing complexity of cellular chemistry and ecology has resulted in the evolution of *problem-solving skills* which form the roots of overt 'intelligence'.

Problem solving skills, which in one form or another may be required for the satisfaction of the motivations touched on above, also have a very ancient heritage. Many relatively simple life forms are capable of feats in this regard that may at first glance appear surprising, but a significant part of this ability lies in memory of one form or another. Here we aren't necessarily envisaging complex memory as we encounter it in higher life forms but at a level and degree of simplicity corresponding to the organisms in question. Memory in this more basic sense is largely based on impressions left by negative or positive experience concerning self-protection, sustenance, and reproduction, which influence future action. This has been observed even in some relatively primitive species, such as slime moulds and certain types of amoebae, which we are accustomed to considering as being simple organisms, possessing no recognisable neurological apparatus. Many plants have also been shown to demonstrate defensive bio-chemical reactions based on past experience - and to communicate that response to other plants of the same species in the vicinity via a variety of biochemical means. Memory as we encounter it in its much more complex forms in higher species including humans nevertheless still has roots in and reflects these same basic organic functions.

Predictive skill based on this interplay between emotional experience/memory and directed action seems central and foundational to the evolution of complex cognition and intelligent behaviour. This clearly isn't an ability that has appeared out of the blue solely to human beings. It has evolved, its origins can be traced to some degree, and the relationship between memory, emotions and action is reflected in the structure of the autonomic nervous system.

The faculty of memory as we know it then, is deeply rooted in our biological ancestry as an aid to survival and the satisfaction of motivation and associated problem solving. Accordingly, the memories underlying thought are rooted in all areas of our experience, sensory and emotional, but this is no longer dictated by biology or genetics alone. As I've touched on and will discuss further, this is because the predominant evolutionary domain in which we now function is one of *shared experience*, which we refer to as culture. The impulses of biology remain fundamental and integral to the purposes of cognition of course, the more ancient layers of cognitive function are still fully present and active, essential in fact - the penthouse is just the top floor of the building, which is otherwise fully occupied and active. Hierarchically though, this sphere of collectively shared experience - human culture - is the principal active domain now, and despite its practical roots, operates in an entirely abstract zone, which is why our attention, individually and collectively, tends to be now dominated by conceptual thought. Nevertheless, the underlying evolutionary imperatives and functions remain much the same as for the amoeba... These two aspects of cognitive function - motivation and memory-based problem solving - can be seen to have evolved entirely intertwined.

Consciousness and the Brain

A further observation is this; all forms of cognition, as we currently understand it, involve electrical exchange between neurons that indicates their activity. Certain patterns of this activity may suggest the presence of consciousness, but generally, the presence of neuron activity only indicates the functioning of particular areas of cognition.

It's apparent that the expression of consciousness in the human organism, or any sentient organism in possession of a nervous system, depends on the brain. However, in the view that I've been discussing, conscious experience is at the 'cutting edge' of cognitive development, which has preceded and been the basis of, the evolution of the brain. In other words, the evolved structure of the human brain is the cumulative *result* of the work of conscious experience over hundreds of millions of years, as well as an instrument of its expression. Clearly, if the immediate under-lying support features fail for any reason - and that may come about through a failure at any level of the brain's structure, molecular or biological - then conscious experience may no longer be possible, but that doesn't necessarily mean that the essential principle of conscious experience is a result of highly developed cognitive functioning. In my view, as I have discussed, evidence of consciousness is apparent in the behaviour of organisms very much less complex than us. What size is the brain of a spider, or a fruit fly...?

A clumsy analogy that comes to mind here is that of carving machinery used to cut and shape all sorts of raw material, great and small. The cutting head will carve wherever and whatever it is directed towards. At the same time, this central in-strument must be sustained to function. It needs to be provided with energy, lubrication, control and so on. Relating this image to the situation of the human mind, clearly the 'cognitive machinery' involved in the act of experience is complex; where can we distin-guish between the 'cutting edge of experience' and the wide range and dimensions of cognitive functioning? Looking a little more closely at the 'cutting edge' of the machinery in my analogy, the actual point of contact between the cutting teeth and the material with which it engages represents the 'contact point' of conscious experience, not any of the structural or control mechanisms involved. In the case of machinery for excavation or carving, such cutting teeth are made of, or coated using, the hardest material available. Such material is inert, and harder than anything else it may encounter which is why it is appropriate for this kind of task. In evolutionary processes this is the 'function' of consciousness. It isn't any of the functions that support it, all of these have evolved over great periods of time and belong to the 'machinery' involved, and to all the mental architecture that has been created in the process, cumulatively, under the directing influence of immediate conscious experience. Just as we use diamond or titanium for their innate qualities, you could say that evolution within the domains of cognition employs conscious experience as the core substance of its 'cutting edges'.

To summarise this point in terms of hierarchy, functions of cognition now direct the course of the evolution of biology and physiology (via the processes of natural selection), and it is by virtue of this relationship that the brain and all its instrumentation has come into being. For that reason, although the brain is physically essential, and is the instrument of expression of consciousness as we know it, nevertheless from this perspective it is a *result* of the constant contribution of conscious experience to the carving of the chambers and dimensions of cognition over evolutionary time, rather than its primary source.

The Conscious and Unconscious Mind

It is generally now recognised that most of the activity of the mind, human or otherwise, takes place at unconscious levels, the conscious level appearing to be something of a bit player. As I have discussed in the last few paragraphs, I'm inclined to see this in the terms that all the evolving 'chambers' of the mind are 'carved out' as a result of conscious experience, a process which has been ongoing over a vast period of time, beginning long before the appearance of mammals, or even dinosaurs, in the remote distant origins of the earliest sentient life forms, around 580 million years ago or earlier. The evolution and developments of conscious cognition at all stages accompany corresponding developments of physiology gross and subtle, and together these are minutely registered through the selection process as genetic inheritance. Viewed in these terms overt conscious experience is at the 'cutting edge' of evolutionary progress and is only minimally required beyond that to support established functions of cognition. This is why many areas of the human psyche, such as the subconscious, function autonomously with little obvious requirement of conscious involvement, and often seem to know more than 'we' do consciously. Current ideas concerning human psychological architecture, such as the id, subconscious, the unconscious etc. were first propounded in times when understanding of evolution and its significance towards our understanding of the human psyche was little developed. The predominant context was that of Western medical science of the time. Consequently, there was a strong link in this form of interpretation with our understanding of the physical body, which rested to a great degree on surgical explorations in the course of treating disease and injury of one sort or another. Evolution theory played little or no part in such explorations historically, nor in the theories in which context psychology was developed, which was that of studying and describing the causes of malfunction of the psyche rather than the nature of its normal functioning - in other words, from a pathological perspective. In a sense the structural descriptions of the psyche provided through this approach can be compared to the anatomical description of the body. Just as awareness of the evolutionary background to the human body provides a fuller picture of the development of our physical form and its relationship to the rest of organic life, so also it provides the broader context in which our understanding of the psyche can be understood. Again, details are important, but context gives a fuller picture.

I've tried to stick to logical arguments in presenting these ideas concerning consciousness, but clearly, I have offered little or nothing new in the way of practical evidence, just a degree of deduction from observation of existing information plus a good degree of speculation. I will only repeat what I said at the start of this piece, which is that consciousness is particularly difficult to define and pin down as a phenomenon. Regarding evidence for the presence and influence of consciousness, there don't appear to be easily identifiable approaches to practical experimentation. Perhaps some of the most intriguing observations that have been made in a scientifically practical way have arisen as an incidental consequence of other inquiries, initially concerning the behaviour of subatomic 'particles' of matter, or energy, referred to as quanta. Here, it has been demonstrated that the very act of observation, by engaging in measurement at that level of events, appears to have a material influence on the nature of particles' behaviour. Even so, this exploration is more focussed on understanding ob-jectively the behaviour of the material world, under the ethos of prediction and control, than on the nature of consciousness, which remains largely in the background of curiosity. It seems to me that however this area is approached, exploring the nature of consciousness is clearly an intriguing and relevant area of further inquiry. If I've succeeded in providing some food for thought, that will do.

A final comment regarding possible further developments in the evolution of cognition and the expression of consciousness is that whatever form it may take, further emergence is always a result of collective dynamics and can't be understood, at any level, purely through the observation of individual entities or factors. A key feature underlying all developments is, again, the principle of *hierarchy*, whereby all particulars must be viewed in the context of their environment.

Summary

Like all concepts that of consciousness has a history. To consider it and its place in evolution, it's necessary to bear in mind some sense of that history. Before we had any understanding of evolution, it used to be thought that consciousness, intelligence, and self-awareness were essentially one and the same, and belonged only to the realm of human experience. This focus on the nature of human mentality is in the background when we now think of consciousness as being emergent from intelligence, but the evidence is that the ability to experience is widespread throughout animate lifeforms. I would say that this should be clear to any observer prepared to recognise and review traditional assumptions about the uniqueness of humankind.

If we take consciousness as meaning the ability to experience, we can consider cognition as representing the dimensions of its expression. Across the wide range of animate life, these dimensions of cognition are many, but essentially, on these terms we can infer the presence of cognition of some form, and therefore consciousness, from the presence of a nervous system, however elementary. We have evidence of this kind extending back in time for around 580 million years, when animate multicellular organisms began to emerge. The human field of awareness comprises several layers of cognition emergent over this time that can be outlined broadly as; sensory and emotional experience, abstract awareness, self-awareness, and conceptual thought. All these layers are present and contribute to our field of awareness.

In this chapter I have discussed the evolution of cognition, as the expression of consciousness, touching on the idea of intelligence without exploring it in detail. In my view consciousness is a more important topic than intelligence in considerations of evolution, because it represents the 'experience of existence' as central to the emergence of cognition, (and therefore 'mind' at all scales), also, because it emerges directly from roots in cellular processes. Exploring its roots of origin must be a matter of great interest...

Chapter Notes

*1 which also begs the question, what's the difference between consciousness, awareness, and attention? Since these terms arise frequently, I'll try to outline my understanding of their meaning. I would say that awareness and attention are intimately related terms, where awareness is effectively equivalent to consciousness, and attention indicates the directing of awareness towards particular features of circumstance either by reflex, or intention, or a mixture of both. In general use, awareness and attention tend to be used somewhat interchangeably, but with awareness as the more passive, general, form, attention the more focussed/active. Each suggests the directing of consciousness/experience towards contexts general or particular, while consciousness itself is the central principle, of both. Using light again as an analogy, if consciousness is compared to light for illumination, awareness would be something along the lines of the general field of illumination, while attention might be compared to the directing of the light towards particular objects or areas. So again, consciousness in these terms is the overall principle of experience, awareness and attention are modes of its expression. This should just be seen as an outline idea of their relationship.

*2 *from this angle, self-awareness is also a matter of degree, arising as a natural result of the evolution of abstract awareness, not of conceptual thought. It may then exist on a spectrum that can be seen to extend much further into the animate domain than has been recognised by current methods of cognitive testing - and to varying degrees in the human sphere - rather than as a binary 'either/or' form of awareness.*

*3 *it seems to me that this view has roots in older, Western theological views of nature and man's place in it. Nature was regarded as subservient to human requirements and could be treated however we wanted. Considerable vestiges off this view still underlie the ethical approach modern biological science adopts towards experimenting on animals and are reflected in the assumption that they aren't 'conscious' and therefore don't really suffer, which is, to my mind, preposterous.*

*4 *this is because we know that the presence of neurons indicates the presence of sensory experience and action - i.e., cognition - of some form. However, to an extent this is something of an arbitrary dividing line, that reflects our ability, and possible limitations of that ability, to recognise and relate to the kind of behaviour that may be involved. It may well be that cognition of other forms exists based on principles which we can't currently recognise with any certainty. I'll stick for the moment to confining my speculations to areas we can be reasonably confident about, viz the nervous system based on neurons, which is present in all complex life forms with motor functions of some degree. Again, these appear to have common origins that go back at least 580 million years and are first identifiable, indirectly, in the fossil record from around that time, as being represented by a simple form of worm. Still today, creatures that we may think of as being simple and lowly - such as worms - contain and express a nervous system based on this pattern - essentially the same basic pattern as our own.*

*5 *this term 'sharing' refers broadly to several different levels of engagement between organisms, all of which could be termed 'communication'. For example, many organisms, including plants, 'share information' via a wide and complex variety of chemical means. In the case of plants this takes place via the air and at the level of the roots. Where free moving organisms are concerned, from insects, up the evolutionary ladder to human beings, this chemical communication is largely represented by pheromones, which play a part in social behaviour at a level that precedes that of overt conscious awareness, i.e., at an established, instinctive level. Within the further evolved sphere, of conscious culture, the meaning of 'sharing experience' naturally relates to more advanced forms of communication. Juvenile members of a population go through extended periods of learning which continue to some degree as they age. On one hand, they learn basic survival skills from their parents, on the other, they learn more broadly from their peers, simply through observation, while contributing in the same way.*

*6 *in addition to these influences on inherited characteristics, i.e., the absorption of experience through social sharing as I've outlined here, and in the well-known terms of direct genetic inheritance, it has recently become clear that variations in the activation or suppression of some genes takes place, triggered by prevailing circumstances. This appears to represent a genetic principle whereby certain beneficial adaptations are stored in the genes archivally in a dormant form and accessed as triggered by circumstances. This is referred to as epigenetic response, and investigation into it is still in early days. Although evidence is largely observable at the level of physical adaptation it's inevitable that cognitive functioning is also involved. It's also probable that there is a corresponding degree of influence on shared behaviour patterns at the collective level.*

*7 *conventionally, it is generally considered that the social and practical constructs of species such as ants arise entirely through the effects of random* *mutation filtered by natural selection over great spans of generation. Such social constructs are therefore viewed as having a quite different basis to those of humans which appear to be less strictly genetically determined, much more a result of abstract thought. But it seems to me that essentially the same kind of process may be at work, with proportional differences in the contributions made by the individual to the form of collective behaviour. Humans appear to have a relatively high degree of adaptive flexibility, while ants appear to have much less, but this appearance may be something of an illusion. Perhaps some external observer from another planet would see our behaviour in much the same way we see that of ants, noting the similarities and regularities of collective behaviour, rather than differences developing from the unique input of individuals...*

*8 incidentally, throughout this outline the focus has been on practicalities, but what about the perception of beauty, a subtle quality of conscious experience traditionally thought to exist only for humans? Resonances can be perceived between the human experience of beauty and the constructive communication between plants and insects that has contributed greatly to the form and diversity of complex life. Take bees and flowers for example; the shapes, colours and scents of flowers have evolved in conjunction with the cognitive sensitivities of the insects that assist in their pollination. The pleasing appeal to human sensitivities suggests to me some very ancient shared ancestry in the cognitive makeup of bees and humans. In the intimate relationship between insects and plants it seems to me that the dominant aspect lies in the cognitive sensitivities of the insects. Beauty, indeed, lies in the eye of the beholder, but has ancient origins and is woven throughout many different features of animate behaviour and the wider organic environment.

*9 are thoughts 'real'? Are the thoughts you are thinking right now 'real'? Do they 'exist'? Clearly, they do, their existence is recognisable just as

validly as the chair you may be sitting on, or the air you breathe, even though they're not physically tangible, can't be weighed, or measured in three (or four) dimensional terms...

*10 another analogy here could be that of electricity powering diverse systems of energy exchange, electronics, and information control, which are at the heart of modern technology. While electricity as an energy source is formless and invisible, it has its own intrinsic qualities that inform the nature of the response of any given electrical/electronic environment/system to its presence. Those qualities are integral to and underlie, the design of the environment's electronic architecture. By analogy, so is the nature of consciousness integral to all structures of cognition; from those of sensory experience and rudimentary memory perhaps, at the most elementary level in animate nature, to those of higher expression such as emotional and rational experience. Perhaps the most obvious difference is that electronic design is deliberately planned while cognitive design progresses under the overall principles and refinements of natural selection, but otherwise the analogy seems reasonable to me. In modern times we take the relationship between electricity and its widespread applications for granted, but most people have only a very vague understanding of their workings. We leave that to those who are more knowledgeable of such matters (actually, I would say that very few people - including those who work with it every day - have any depth of understanding of electricity beyond that based on its practical applications). Nevertheless, the development of that understanding has been enough to make a multitude of processes and phenomena in nature comprehensible and accessible to discovery and invention. I believe that the same applies to the relationship between consciousness and cognition, in gaining some understanding of the evolutionary sphere of 'mind'. One final observation I would draw from this analogy is this; to deduce that consciousness emerges from complex cognition is like deducing that electricity emerges from complex circuitry...

Chapter 6

Intelligence

Outline

I've discussed the idea of intelligence to some degree in considering consciousness Although these topics are intimately related, I've given them separate chapters in an effort to distinguish what we mean by them. It seems clear to me already from the considerations so far, that intelligence as we encounter it is an emergent phenomenon, widely expressed in nature, far beyond the limitations of our traditional views of it as unique to human beings. Nevertheless, because it still represents a significant aspect of current thinking on the relationship between the human condition and life generally, it feels necessary to give it more attention. so, I focus on it a little more here.

What does 'intelligence' mean?

Again, I would remind the reader that the way we use language has to be borne in mind. First, I would observe that the word 'intelligence' is a noun, but relates to a *quality of behaviour*, which is non-static, and is better conceived of in terms of activity in which the quality 'intelligence' is perceived. When we conceive of it as a noun, the tendency is to look for an embodiment, or personalisation, of that quality. Accordingly, the general notion of 'intelligence' incorporates the perception of deliberate, considered action. When we observe behaviour which appears to express such qualities, we tend to infer the presence of deliberate and independent mind/s, which we then characterise as 'intelligent'.

To a great degree, our views of intelligence reflect perceptions and assumptions drawn from human behaviour. Until relatively recently, and still now to some degree, in the traditional Western view, the faculty of intellect has long been regarded as being the seat of intelligence and present only in humans. Even so, it has always been apparent that life expresses a high level of coherence of a kind that often suggests 'intelligent' organisation in many different spheres, from the complexity of plant life to those of ant and bee colonies as well as many other animal communities (not to mention the wonders revealed through microbiology). In that wider context, intelligence is used in reference to organised processes and activities clearly not dependent on the exercise of intellect. One only has to observe the behaviour of many animals such as cats, dogs, birds etc. to see the expression of intelligence in their behaviour, without necessarily the presence of intellect in the human sense of an instrument dedicated to abstract reasoning. Detailed examination of the methods of organisation expressed in living processes, through the studies of evolution and genetics have revealed unambiguously that the patterns and organisation of life, far from resulting from any exercise of intellect, develop spontaneously and coherently within their environmental circumstances. In these conditions the notion of intelligence as essentially a function of intellect needs reappraisal. When we speak of 'intelligence' then, it seems to me that we derive our use of the term mainly from our human perspective regarding behaviour generally. One of the very influential elements of this view is the perception of 'intentionality', or 'deliberate action'. I don't intend to get into a long discussion of all the different views that exist on the topic of 'intention' in nature, I'll simply offer my view, in the context of evolution.

Intentional Action

'Intention' implies forethought and prediction, which in turn indicates motivation and observation, i.e., directed awareness, interpretation, and experience-based memory. A broad sense of this disposition is reflected in the word 'planning'.

The ability to observe, interpret, and act accordingly with forethought, is characteristic of the faculty we call 'abstract thought' which we once regarded as being the sole preserve of the human intellect. We now know that this is an ability that developed over long periods of evolutionary time. It didn't appear out of the blue in homo sapiens but is certainly expressed also in many other species, though perhaps to a lesser degree. It is worth considering the notion of *forethought* a little more closely to recognise and understand this point. I would say that our usual interpretation of the meaning of forethought reflects our impression of human thought and planning, which is complex, involves a considerable conceptual element, and doesn't appear to be represented elsewhere in nature to a great degree. This human faculty is what is usually referred to as intellect, and its exercise is the basis of our traditional meaning of 'intelligence'. Behaviour in animate life forms that doesn't appear to correspond to this faculty is traditionally referred to as instinctive and is not generally regarded as 'intelligent'.

Actually though, as becomes clear if one observes one's own thought processes just a little, much of our faculty of forethought doesn't involve conceptual thought at all but still fully engages our skills of prediction at a more basic level of abstract awareness, and these same skills are widely represented in nature,

(I have made this observation already, so this is something of a repetition, but the topic of 'intentionality' makes it worth revisiting). The main practical difference is in the time scale, complexity and extent of prediction involved. I'll use again the image of a bird in flight to present the observation that this activity involves established instinctive aspects but also requires complex flexible cognitive skills of evaluation and prediction of the environment - wind conditions, take-off, and landing, tracking prey etc. For a bird, its decisions are therefore based on a mixture of reflex/instinctive motivation and motor skills, and short-term evaluation and prediction based on acquired experience - i.e., memory-based learning. At root, we humans also use cognitive faculties like these all the time, and this to me is at the heart of what we refer to as intentional behaviour. This is essentially 'predictive', but on a short-term scale and without the requirement of conceptual thought/planning. The effect of conceptual thought is to extend considerably the range and scale of this predictive ability in time and space, but the basic principle of 'intentionality' remains essentially the same. In these terms, 'intentionality' and its expression in 'purposeful' human behaviour, *including* conceptual thought, is simply an evolved extension of this functionally purposeful behaviour expressed widely throughout animate nature. This level of abstraction has however reached the stage where we have come to perceive ourselves to be in possession of a faculty of intentionality distinct from that of the rest of nature. I would say this distinction is something of an illusion that tends to blind us to recognising that *purposeful* action is a much more basic and widely expressed feature of animate nature. Let's take a closer look therefore at the terms 'intention' and 'purpose' and, if possible, make a useful distinction between them, from an evolutionary perspective at least...

At the level of the chemistry of cell functions, the molecular activity involved appears indistinguishable from inanimate chemistry other than in terms of complexity. In any context, atoms don't have to consider the future to execute their functions in combining to form molecules. They simply act within their given environment according to the inherent principles that underlie and govern their behaviour. As observers we wouldn't perceive this to be either purposeful or intentional. Motivation here is just the compulsion of atomic forces that we refer to purely in terms of physical principles. At the level of even the simplest single cell organisms, the requirements of cell survival and propagation are highly complex, nevertheless their internal chemistry appears to represent a natural extension of the 'blind' chemistry involved at the inanimate level. In terms of that chemistry, we could say that they are motivated to survive and propagate, but 'motivation' at this level still involves no form of planning or forethought, no intentionality. We might say therefore that their actions *appear* dedicated to specific ends - 'purposeful' - as a reflection of the complex and very particular requirements of their continued existence but are no more 'intentional' than those of atoms or molecules. In practical terms, the use of the term 'purposeful' here really reflects the view of the observer in interpreting events rather than any acts of 'intention' on the part of these organisms and the complex self-sustaining organisation represented by even the most elementary of living processes. As observers therefore, we perceive behaviour that represents a degree of coherent molecular organisation which appears 'purposeful' beyond the scope of strictly inanimate natural processes, but doesn't appear to represent any expression of 'intention' as such.

In these terms, 'purposefulness' is perhaps a more subjective term than 'intentionality', referring to the observer's perception of coherent behaviour, distinct from that of purely physical principles, but without the expression of prediction necessarily. This represents a higher level of self-organisation than that which is observable at the level of inanimate chemistry, but which we would not perceive as expressing intentionality. 'Intentionality' necessarily implies prediction, which only becomes a factor once cognition that allows this in some form is present. One might say that both 'purposefulness' and 'intentionality' are conceived of as emergent qualities, 'purposefulness' emerging from a background of molecular principles, 'intentionality' as further emergent from the sphere of purposeful self-organisation (another way of putting this is that, in English, all words that end in 'ly' are adverbs but not all adverbs end in 'ly'. Intentionality is a form of purposefulness - but not all purposefulness is intentional...). As ever, this distinction can be best understood in terms of hierarchy. Still, it should be borne in mind that these are terms of observation only, intended to assist in interpreting a continuity between apparently quite different types of behaviour in nature. Viewed in this way, motivation can be conceived of as a spectrum, with the 'mechanical' forces of inanimate nature at one end and conscious intention at the other. As I've mentioned, evidence is very strong that all life as we know it now can be traced back to a common ancestral cell (LUCA). Although this is as far back as we can deduce currently with some reasonable degree of certainty, this ancestral cell would not have appeared suddenly overnight but would most certainly have been the result of much prior 'experimentation' on the part of molecular chemistry (discussed in Chapter 4). A significant point here is that the appearance of distinct molecular entities - the earliest proto-cells - may represent the point at which partially isolated and sustainable complex chemistry became possible, as they provide the condition of limited isolation from the environment on which all known cell metabolism depends. In terms of evolution, this may represent the point at which blind inanimate forces that drive events at the molecular level *began* to translate into the more focused sphere of complex organisation we might characterise as 'purposeful'.

'Intentionality' on the other hand, suggests the further extension of the drive of 'purpose' from its roots in immediate, blind, chemistry to the *anticipation and directing of its satisfaction* through action. This implies cognition and therefore consciousness. Close examination over the last several decades in the sciences of biochemistry and other disciplines suggests that the transition from inanimate chemistry to early cellular life was one of gradual emergence, rather than some sudden overnight single event. Similarly, it would probably be a mistake to think that there is some sudden point of departure between the condition of 'blind purposefulness' and 'intentionality'.

In summary, viewed in these terms it could be said that 'purposeful' activity can be understood as being emergent from molecular chemistry, 'intentional' action as emergent from purposefulness, bearing the particular feature of *prediction*. If we consider the actions involved in elementary cell metabolism as the base line of organic organisation, we can say that cell behaviour is coherent and 'purposeful' in terms of metabolic processes, but with no contribution from 'intentionality' in the sense of prediction of any form. In terms of cellular function, 'intentionality' is entirely absent, but 'purposeful activity' is fundamental to all organic life. 'Intentionality' is only identifiable at the level where faculties of prediction have emerged. We traditionally associate this capability with the human condition, but we now know that although we may represent its most complex development so far, it is clearly also represented much more widely in higher life forms. Therefore, although in terms of human behaviour we tend to view *purposefulness* and *intentionality* as being the same thing, from an evolutionary perspective it would be more useful to distinguish our use of them to some degree. In my use of these terms here, purposefulness is used to represent coherent motivation at all levels of organisation in organic nature. Intentionality can only be attributed to behaviour based on the exercise of prediction, but this is by no means limited to the sphere of human behaviour.*1

I would emphasise here that none of these concepts I have been using, such as intelligence, purpose and intentionality represent definitive, objectively existing qualities, all are concepts of description we create and are free to explore. If anything, my purpose is not so much trying to define them, but to explore them and encourage the reader to do the same.

Chapter 5 presented a view of cognitive structures based on the relationship between distinct but integrated features built up over long periods of evolutionary time. In this view, immediate conscious experience, in the present, is in the evolutionary 'driving seat', while most of the established functions, have been laid down over the distant past and function in a mostly autonomous manner. An analogy that may help illustrate this situation is that of an ocean-going ship with many areas of activity and crew members that work in close co-operation. At the extremities of the tasks that need to be dealt with, these are carried out by operatives whose competence has been established, tried, and tested to the degree that they can be relied on to function autonomously with a minimum of supervision. Hierarchically positioned above these are minor officers who supervise them but also carry out more strategic and flexible functions, whose competence must still be relied on, but who require more supervision and direction to ensure smooth synchronisation between operational 'departments'. This supervision and associated direction come from higher still up the hierarchal chain. At the highest point of this chain of supervision and command sits the 'captain' who makes overall decisions, studies the environment, navigates the ship's course, and directs the overall course of action but does little, if any, of the actual functional work of the administration. This position nevertheless represents the seat of 'intentionality' and prediction, perhaps the most important choice and decision-making point of the entire structure.

The main task of this 'captain' is not to be aware or 'conscious' of every detail of the ship's operating systems - they work best with minimum interference - but to access and employ its vast store of information and skills to make decisions of an overall higher order kind concerning navigation. The central faculty the captain exercises in decision-making is judgement, or *discrimination* (which I'll discuss in Chapter 7). Part of the helpful symbolism here is that the captain is usually positioned where they can have a good view of the environment being navigated, has direct access to steering and related controls and, ideally, well organised supervision of all other functions. Basically, he/she is the one in a position to experience the overall environment, inner and outer, and make crucial decisions accordingly. The rest of the ship's operations are purposeful and supportive of that. Intentionality - active, conscious decision-making, resides in the hands of the 'captain'.*² Again, this illustrates the distinction between 'purpose' and 'intention' as one of hierarchy.

Mostly then, the higher order choices which must be made are presented to the 'captain', the conscious mind, by the unconscious mind, though the 'captain' can and continuously does, actively access information from the various 'departments', from which choices and decisions emerge. So, the process is interactive between the 'captain' and the departments - that is, between conscious intention and unconscious or subconscious processes - but the hierarchy is maintained. In this analogy it's easy to see the contrast between the great complexity and sophistication of the technology and organisation required in such a vessel, and the comparatively diminutive contributions made by the captain controlling it. But coherent navigation is crucial to the safety and progress of any such vessel, as is well recognised. Nevertheless, this pales into insignificance when we consider the incredible complexity and sophistication of the human body, mind and nervous system, and the importance of its 'captain', conscious mind.

The problem that the conventional notion of intelligence presents then is rooted in the associations of this concept with deliberate pre-planned or designed action weighed according to human models of intentional behaviour. Evidence we now have tells us that the form of human intelligence, and the faculty of intellect, is a natural extension of simpler forms of predictive cognition, widely present in living nature, emerging very early in the evolution of complex life-forms, and with roots in organic nature, ultimately in the 'purposes' of cell metabolism. In some ways this perceptual adjustment is akin to that of recognising that the earth isn't at the centre of the physical universe - that the concept of intelligence as we conventionally use it is something of an anthropocentric relic. This doesn't do damage to the notion of intelligence as such, on the contrary it acknowledges that this aspect of nature is a ubiquitous feature of complex life, and that human intelligence is just a particularly developed form of it that nevertheless still reflects its origins, being directly physically and mentally on a continuum with the entire history of its evolution. Another observation I would like to make here is to point out that the faculties of discrimination and logic that contribute to intentionality are also innate features of our mental makeup; they aren't separate in any way from the autonomous features of our organism, they are also autonomous features. The perception of intelligence as necessarily an expression of 'individual control' is an illusory result of self-awareness.

Having made the point about the ubiquity of 'intelligence' throughout animate nature it seems reasonable to observe also that its expression varies greatly in complexity. A dog does appear to be much more 'intelligent' than a slug, a chimpanzee than a frog. At a cellular level all these creatures share an essentially identical degree of complex eukaryotic form, so their 'behavioural intelligence' is a function of the next degree of evolutionary hierarchy. In other words, the development of expressed 'intelligence' is a feature of evolving complexity. Again, this may appear to be rather obvious. What I am attempting to do here is to put things into a coherent perspective and that requires putting them in the right order and relationship. In the same way that organic life governs molecular processes, the domain of cognition governs the direction of the evolution of cellular complexity, rather than the other way round.

Over these two chapters I have offered the view that 'consciousness' is effectively a core principle in cognitive evolution, while 'intelligent behaviour' is an immediately linked and constantly evolving result. In this view the essential principle of consciousness itself doesn't change, but all of its means and dimensions of expression do, and this includes all forms of 'intelligent' behaviour to be observed in living creatures, individually and collectively.

I briefly refer again here to the analogy of electricity in an electronic circuit. This is invisible to an observer but essential to its functioning and revealed only through its effects. The 'circuit' in this image, represents the entire domain of cognitive evolution and its degrees of complexity; like electricity, 'conscious experience', remaining formless, nevertheless, under the domains of cognition, provides the ingredient central to and underlying all that transpires in evolutionary form, physically and mentally.

Summary

From the perspective I have been presenting over these last two chapters, consciousness and intelligence are inseparably entwined phenomena, where intelligence as we encounter it, is a result of the expression and evolution of conscious experience, rather than the reverse.

There has long been a tendency to equate intelligence with rationality - and implicitly, intentionality. This may seem reasonable at face value, but this use of the concept intelligence is very partial and limited in its ability to represent the wider dimensions of conscious experience involved. These need to be recognised also as entirely equal in significance to rationality per se in providing a fuller picture of the meaning of 'intelligence' as they are complementary and inextricably related. This isn't a new observation of course, it has long been recognised that a human is more than just feelings and more than just a rational brain viz the term, human 'being'.

Similarly, when we consider for example, the behaviour of dogs or cats or many other animals, the qualities which we recognise as expressing 'intelligence' plainly include emotional behaviour at least as much as the ability to reason. Most of this topic was covered in the previous chapter, on consciousness. In this chapter I focused on the perception of 'purposefulness' and 'intentionality', their roots in nature and their contribution to the perception of intelligent behaviour.

Chapter Notes

*1 in considering the distinction between purposeful behaviour and intentionality there may be few more interesting examples of coherent organised behaviour than that offered by some species of slime mould. These exist primarily as independent amoebic organisms but can coordinate their activities to behave as colonies, which to a remarkable degree resemble organisms of a much larger scale, in behaviour and appearance. This integrated collective behaviour takes place particularly when food is scarce and relocation is necessary, and when reproduction is required. How these collective activities are achieved is little understood, but it seems to be a clear expression of 'purposeful' behaviour in response to conditions in the environment, but without the presence of 'intention' in the sense of prediction.

*2 it has become apparent that many decisions are already made subcon-

sciously before the conscious mind takes over. Some would interpret that as indicating that the sense of there being 'a' captain in control is no more than an illusion. But it seems to me that it just represents the captain accepting the 'advice' from experienced 'departments', with a minimum of active involvement.

Part 2

Throughout Part 1, the focus of attention has been on the physical background of evolutionary processes, culminating in the emergence of conscious cognition.

In Part 2, the central theme concerns ongoing evolution within the sphere of human culture, as representing the most advanced domain of evolutionary complexity that we know of currently. This is considered in terms of the same basic principles of evolution observable throughout organic life which have been discussed so far.

The relationship between the 'individual' and the 'environment' is seen to be at the heart of ongoing developments, and is explored here. The dynamics of this relationship are primarily those of conscious experience and for that reason the first chapter in this section explores some aspects of the ways in which we interpret and describe experience.

I would like to remind the reader that my descriptive style throughout this work isn't intended to be viewed as statements of 'facts' per se, but rather, as those of an observer, like a gardener perhaps, digging and describing what I find or encounter as I proceed...

Chapter 7

The Observer

"There is a crack in everything, that's how the light gets in."
(Leonard Cohen)

Outline

It doesn't really require more than a few moments reflection to recognise that the world we inhabit has been hugely influenced, directly and indirectly by the inquiry and discovery that has taken place over the last few centuries in the fields of the sciences. Along with the vast technological changes that have resulted, there has been a corresponding upheaval to the dominance of older, largely theologically based, views of the world, nature, and the place of man. Nevertheless, it is also clear that the knowledge structures on which the sciences rest - and all related and influenced philosophical views - are also in a constant state of flux and ongoing development. Simply, like everything else, they evolve.

However substantial reality may or may not be, our experience and attitudes greatly influence our view of 'it'. There's nothing new about that observation and recognition of this lies at the heart of the scientific ethos of objectivity, that is, the constant effort to distinguish 'objective' reality from our necessarily partial, or 'subjective' view of it. Accordingly, at least in the public mind, scientific theory tends to be heavily identified with the attempt to establish irrefutable 'factual' knowledge, while aiming to be open-minded and free from subjective distortion. Clearly, the position of the 'observer' is a crucial factor in this endeavour. Here I explore this topic a little.

But what, one may ask, is the relevance of this to the subject of evolution? Well, there are two points of relevance. The first is that the overall view of evolution I am presenting here is not entirely conventional, and to get a handle on it it's necessary to look behind some of the established pillars of conventional thinking and be particularly aware of the importance of the position of the observer. Whether you come thereby to agree with the views I'll be offering, or to disagree - and I'm sure there will be a good mixture of each - the very act of thinking about them should prove interesting and useful.

The second point is intimately linked to the first, and this is, that the way in which we use language is central in the formulation and communication of all ideas, and the understanding of that isn't to be found in books of philosophy, science, or religion (or dictionaries), but rather, by looking in the first place at how we, personally, use it.

The concept of evolution is, of course, normally used in reference to events that can be observed and interpreted, but it is now well recognized that the position of the observer predisposes the results of all observation and the background to this position also evolves over time as our knowledge and understanding develops. So, it's necessary to pay close attention to factors that contribute to our frames of reference as a pre-requisite for arriving at views which aren't simply extensions of existing assumptions. In this chapter therefore I try to prelude further discussion of evolution by asking the reader to consider and be aware, to at least some degree, of how we use language to construct descriptions of the world. This requires taking a look at our personal understanding of some concepts which we may use every day but which mostly we take for granted and rarely consider closely. To use concepts most effectively, as with any tools, we need a sense of their strengths and limitations and keep them in good condition. To do that we have to be prepared to examine what they mean to us personally, not just according to established definitions. The meaning of concepts exists by collective consent, and this is flexible and organic - they don't originate in dictionaries but in living use (to later be collected and deposited in dictionaries, you might say, like butterflies or flowers...). When we recognise this, we can make more meaningful and constructive use of them.

Systems of knowledge also evolve, along with everything else. Generally, though, when we talk of evolution in this kind of context we aren't really thinking of natural selection, competition and so on, just the development of ideas and information over time. Nevertheless, the same kind of factors of evolution being considered so far do come into it. This chapter considers some basic features of the way knowledge is formed, and how this acts as a lens, or filter, through which we view the world.

It may be difficult to see the relevance of some of this content at first, but I would ask the reader to bear with me because it is an important step in presenting these ideas. Here I try to prepare the reader for the discussions that follow by looking, in the first instance, at some general conditions that underlie the formation of ideas/theories, as I understand it. To introduce the ideas that follow concerning evolution, it's necessary to begin by having a close look at how we tend to think about things based on many established assumptions, and set the stage for the reader to be aware of their own, to at least some degree... This chapter may be regarded as 'preparing the ground' for what follows. As ever, these observations represent only my own view of things, to be considered...

Knowledge

Usually when we talk of knowledge the context concerns human behaviour, but clearly most other creatures possess certain forms of 'knowledge' also. Knowledge you could say, has layers. A spider 'knows' how to spin a web, a rabbit, how to dig a burrow. This kind of knowledge is instinctive, that is, genetically established. Then there is knowledge acquired through practical experience, such as where to find food or safety for example, which we call memory. In such cases both forms of knowledge function together; innate knowledge and acquired knowledge. (Even relatively simple organisms, such as, for example, slime moulds, demonstrate the ability to 'learn' by experience the shortest route to a food source.) Beyond this, at the level of culture (for more complex organisms) knowledge is also acquired through sharing, which represents, in effect, collective memory. All these types of knowledge, up to and including the level of higher evolved species, involve experience of a direct sensory kind that doesn't require conceptual thought. Conceptual thought however, is a form of knowledge creation, description and acquisition through sharing which has emerged more recently in evolutionary terms and *appears* to be unique to humankind. Again, as in all evolutionary structures, the earlier forms remain integral, in this case, with and within, conceptual thought. There is a qualitative distinction between these layers, but there is also structural continuity to them that reflects the central principle of experience of existence which I've discussed in the chapter on consciousness.

Amongst the mental qualities that contribute to the acquisition of knowledge in the sphere of conceptual thought is the faculty of 'reason'. This term represents the application of the perception of logic to events to establish 'truth' or 'fact'. This is an innate faculty that to varying degrees we use constantly in all kinds of circumstance. It's by no means unique to human behaviour, what is unique is the degree to which we exercise it in the construction of complex abstract thought. In contemporary scientific inquiry, reasoning is reinforced (ideally) using careful experimental protocols to produce results as free as possible from ambiguity and tested by independent experimentation and review to establish collective agreement. Philosophically this approach is described as 'objective' in character because it attempts to establish theory and understanding of principles that as closely as possible reflect an 'actuality' or 'reality' considered to exist independently of any particular observer/s. The well-known archenemy of 'objectivity' is of course 'subjectivity', i.e., reliance on, or being over-influenced by, one's own personal view or interpretation of events.

Regarding the collective sphere of shared knowledge, it's worth noticing that the principles on which this works are represented and recognisable in one's own personal experience, it doesn't require any special training to gain a little insight...

In everyday situations, it seems to me, our personal understanding of 'reality' is constantly being checked in our flow of awareness, against our sense of established collective agreement about it; for example, I may never have been to Australia, but I am secure in certainty that it exists because of my contact with many people who have been to or come from there, and by many other kinds of indirect supporting evidence. Expanding and generalising on this observation, I can see that while much of my view of the world and 'reality' is based upon and supported by everyday direct experience, my broader field of view depends on a complex background of *inferred* truth, derived from shared experience and other indirect evidence, much of which is conceptual in nature. This applies to some degree to most of our areas and types of 'knowledge', from that of our immediate domestic circumstances to the finest areas of scientific inquiry.

Experience, Interpretation and Description

Currently, the conventions of Western thought are heavily weighted towards the 'objective reality' of 'matter' with little regard to the factor of 'mind' other than as a potential source of subjective distortion. I find this a deeply interesting and relevant topic and so I will make here some general observations about where 'mind' meets 'matter'.

Which is the more significant aspect of 'reality', matter or the consciousness that experiences it? When I see, hear, feel, smell, or taste something, although I may identify the source of the experience as being *'external' to me* - and for me this is an astounding recognition - clearly the entirety of the experience, including the perception of its source being external to me (and all conceptual structuring), exists *within* my mind. The distinction between 'external' and 'internal' is created at an internal/instinctive level as an intrinsic aspect of perception, and therefore, thought. Still, it takes only a little reflection to recognise that *all* of our experience takes place *within* our mind. This same faculty of 'object' perception is at the root of the sense of our own existence as an entity, or 'object' i.e., just as we perceive an outer field of objects of attention, we experience our own existence as an object within that field; we are self-aware.

Knowledge of reality in any sense begins with direct experience itself, mediated and given structure by the systems of perception. Abstract thought then provides us with *interpretations* of such experience, ranging from those called on by the practicalities of everyday circumstance to those of imagination, theory, speculation, and fantasy. The dimensions of experience reflect, in the first instance, the physical spheres of the functions of the senses and the information that can be derived from them, and this is at the root of how we perceive reality. For example, our faculty of sight and ability to distinguish light, dark, colour etc. is based on our physical response to light, while hearing has its basis in pressure variations in air or other mediums. This sensory information is then interpreted using higher systems of cognition. Together, these represent the basis of perception. Reality then, in terms of both what exists, and how we experience it, is necessarily understood through the medium, and in terms of, the instrumentality of consciousness, which is what the terms 'cognition' and 'perception' relate to. However, while it is well recognized that perception is physically rooted in the mechanics of the senses, with extensive systems of neural processing participating in its structuring, the interpretation of experience also involves considerable moulding by collective, cultural, influences. (Drawing on an earlier analogy, you could view this in terms of 'hardware' and sensory equipment, provided by nature complete with 'operating system software', while the environment provides the conditions, the 'application' level in a flexible and ongoing form.) It seems to me reasonable to conclude that consciousness and the functions of perception, can't be ignored in any consideration of reality, physical or metaphysical, objective or subjective.

The prevailing ethos of modern science rests largely on the assumption that reality exists entirely independently of our experience of it, and investigation of its structure requires the exclusion as far as possible, of the conscious dimensions of experience, in the search for material, 'objective' definitions. From this perspective consciousness, when it is considered, is generally regarded largely as a fortuitous and circumstantial outcome of purely 'physical' processes. However, all inquiry is necessarily carried out through conscious observation and exploration within the very particular constructs and parameters of human perception - both 'hardware' and 'software' - and these condition the entire form all objective inquiry and description takes. To have any valid understanding of reality, objective or otherwise, it is necessary to be aware, to at least some degree, that these practicalities of perception underlie all our ideas, deductions, and assumptions. In the following exploration I'm less interested in the biological 'mechanics' of experience involved as these have been, and are being, extensively studied, than in the filters of knowledge acquired from our cultural background and circumstances, the 'application' level, which are much more flexible.

Language

Language is our main medium of communication. This may seem absurdly obvious. But sometimes the obvious is so obvious that we can be oblivious to its full significance. Language is to social communication for humans as water is to swimming, for fish. The entire range and dimensions of our personal and collective interpretation and description of experience is formed and communicated through language of some form. Also, just a moment of reflection should make it clear that language continuously grows and develops over time, affected by and affecting our views of the world in every sphere. In short, it evolves in conjunction with and reflects, our social environments.

Essentially, language is structured with elements of meaning, the main components of which may be compared to bricks, which we call 'concepts', much of the rest being guiding elements that we refer to as grammar which provide dimensions of relationship for the concepts. Together these may be very roughly compared to bricks and mortar. In addition to the 'practical' information content, language also carries a great deal of emotional content, particularly through intonation, and the use of body language. Together, all these elements create the 'architecture' of language. This outline description is based on my own personal understanding and use of it rather than any formal view. I make the assumption that most people can relate to this description to some degree, though any linguist reading it may not be too impressed...

Concepts

For me, concepts play a role in our view of the world that can't be overstated. In some respects, the points I will try to make here are a key to everything else in this book.

The meaning I prefer for 'concept' derives from the verb, 'to conceive' in the sense of initiating life. All concepts are mental creations, which once brought into existence occupy a place in our shared 'world of description' of experience. I consider that it's essential to recognise that concepts are descriptive representations of reality, not reality itself. All concepts are relative, none are absolute. Many people - including scientists and philosophers - fail to recognise this and search for 'absolutes' of some kind within the realm of description. It seems important to me to keep a sense of distinction between the 'reality' of direct experience and the descriptions we create.

Some of the areas of discussion which I am about to embark on are more usually considered to be matters of philosophy rather than practicality, but it seems to me that these perspectives on knowledge are inseparably intertwined, and it is an error to think that they can be considered in complete isolation. As I indicated in the introduction, I am not a philosopher except in the sense that every person is their own philosopher and, I believe, has the right - and the responsibility - to exercise the judgement of truth for him/herself. So, I gently resist the influence of the weight of established traditions and descriptions to conceive of, and express, my own personal impressions. For the most part these won't be found in any textbook or dictionary. May the reader make up their own mind about their accuracy or relevance.

Direct experience is the root source of knowledge. Conceptualization is the principal instrument we (humans) use for the description, interpretation, and communication thereof, generally in an everyday sense, and in all fields of inquiry. In any inquiry we have to take a close look at the principal terms and concepts involved, as we must first have some reflection on the terms we are employing. In this case, for reasons that should be becoming clearer, some fundamental ones are knowledge, fact, belief, subjectivity, objectivity, reality, and truth.

Primarily, the meaning we have for words is acquired from our environment, which is where we first encounter them of course, but it can also be said that we have our own personal understanding of them, which generally connects quite well with that of other people, but sometimes much less so and in any case rarely examined closely. This is particularly the case where more abstract concepts are concerned. Closer examination can often expose considerable underlying differences in their meaning to us, particularly where they relate to more subtle aspects of our experience or understanding. Interpretation here is often quite vague, and to use such concepts effectively it's important to inquire into how we use them, to have at least a little more clarity about what we mean by them. It seems to me that language is a living instrument of collective communication, with meanings and definitions conferred at the moment of use, created and garnered through the power of collective experience, and established, or confirmed, by convention - very 'organic' in fact. The point of these observations is that the meaning of words, concepts, has to be found in the first instance, by considering, and perhaps refining, our own personal understanding of them, rather than just looking 'outwards' to established definitions. After all, when I say something to someone else, ideally at least *I* should have a reasonable idea of what I am saying and what I mean, otherwise, I don't know what I'm talking about, literally.

As I touched on at the beginning of this piece, it should be recognized that all concepts exist within relative frames of reference. No concept has an absolute value or meaning. They are created by a particular faculty of the human mind to grasp, understand, explain, and communicate aspects of perceived reality, in terms that for the most part, are rooted in, and sometimes extend or modify, concepts already established at any given time and place. So, we might say that the very concepts we use to describe our experience of the world develop, or evolve, over time.

It's likely that there are deeper parts of the psyche where underlying structures of cognition predispose much of the form that conception takes, and this includes its essentially relative nature. It's important to recognize this for various reasons, the most general one being, again, that like any tool, to be most effective, we need to have some understanding of how it works. More particularly, we can realize, with the benefit of historical hindsight that while many concepts we now take for granted originated and were useful and relevant at times and in cultures where certain kinds of inquiry were taking place, many of these may no longer be accurate or relevant in these times and need to be re-considered, because our perspective of understanding has shifted a great deal. In particular, the deep exploration of the physical sciences and the emergence of understanding of some processes of evolution have contributed to this situation. Some older forms of description can benefit from being re-visited and perhaps brought up to date, some may be discarded as no longer relevant; sometimes we need to find entirely new ones.

Many older concepts and views are deeply interwoven still into our mental interpretations of the world we experience (including ourselves) and take for granted to some degree, including some basic ones that I have touched on, such as intelligence, existence, and reality, as well as a multitude of others. Again, a moment's reflection will reveal that while we use such concepts freely in everyday language, no two people are likely to agree completely if asked, on the meanings of any of them. Context largely governs meaning, and in actual use therefore most concepts tend to be used with some flexibility rather than being held to absolute definition. This flexibility is an inherent quality of language, reflecting its roots in living culture and exchange.

Concepts comprise entirely abstract mental frameworks of description - labels, and avenues of association we use to establish and manipulate links between experiences, both of a direct sensory kind and of the entirely non-physical kind entailed within thought itself. The whole field of conception then, is a relative, collectively shared one (though, importantly, all operations of interpretation and shaping take place in individual minds). More generally of course, thought is much more of a 'multi-media' process, involving as it does imagery, memory, imagination, emotion, and rationality, all contributing along with the conceptual aspect to create the world of inner experience. It can appear to us that conceptual thought is a fundamental characteristic of human mentality, but it has really been growing over time to play an ever-increasing role in our engagement with the world at large. As I have discussed in the chapter on consciousness, the faculty of conceptual thought has emerged from a background of a more general faculty of abstract thought that is widely expressed throughout animate nature, and the extent of its use appears to vary between and within different societies, and even individuals.

Direct experience, again, is the central 'act' of consciousness by which we cognize existence and is also the root or touchstone of conceptualization. What I mean by this is that the sense of validity, which we use constantly to gauge the exercise of conception, at all degrees of abstraction, derives from concepts formed initially at the most immediate proximity to direct experience. Terms such as chair, table, person, house etc., are examples of such 'primary' concepts. Concepts of immediate association, such as these, then stretch into broader, more obviously abstract areas, such as home, livelihood, economy, community etc. This is also reflected in how we can perceive and handle numbers we associate with small quantities of objects, such as one, two, three and four, easily as they relate closely to direct experience, even though they are entirely abstract. We then extend these abstractions into more elaborate formulations, which carry their implicit validity, even though much further removed from the immediate proximity of direct experience. In the example of primary numbers, that root validation provides the foundation for arithmetic and mathematics. Now all this may seem reasonable enough but not so obvious perhaps is that the entire field of conceptualization is really *equally* abstract, regardless of its proximity to direct experience, in the same way that an architect's blueprints are abstract, however accurately or inaccurately, grossly, or subtly, they portray 'reality'. No matter how simple or complex the subject matter is, 'reality' is always out of reach of the description. No matter how detailed and accurate a representation those drawings may be, they would never be mistaken for the bricks and mortar, wires, pipes, and woodwork of a real construction. Another comparison I would make here is with the difference between an object and a photograph of that object. The object is 3-dimensional but the photographic image, no matter how good, is 2-dimensional. As such it is a representation from a particular angle of view. There may well be many other angles of view, which will reveal more, and often quite different, information about that object, of equal validity (and limitation). Even if many such images were gathered to provide as much information as possible, the most they could provide is the illusion of 3-dimensionality. Of course, when we are dealing with photographic images per se this distinction is usually obvious. There may well be some strong emotional impact or identification generated, but the image is never, or rarely, mistaken for reality. Where concepts are concerned however there can be a much greater tendency to identify them as reality and a failure to recognise their partiality, regardless of their accuracy.

This abstract, relative nature of conception is important to recognize because all forms of description and theory are constructed using it. When representations are accurate, they can be very effective and useful - templates of reality you could say; it's easy therefore to identify theory as 'truth' or 'fact' when it is never more than a description and *always* relative. Clearly some descriptions provide more accurate representations than others and therefore may be regarded as 'true' but still only 'relatively' true, as description, not true in an absolute sense. For a simple example, one might assemble pieces of wood in a certain manner and call the result a 'chair', but it is the context of its use that validates this name and its meaning. Someone with no experience of chairs may see it as simply a wooden ornament or even just as firewood... For the most part, new experience is rapidly interpreted and categorized within available existing conceptual framework, and this framework grows and gradually, sometimes rapidly, shifts to accommodate new perception or understanding. This process is a dynamic, complex, and subtle one, quite organic you could say; however, the entire field of conceptualization is structured around such relative frameworks and as such is largely bound by those frameworks and can't be used directly as a medium for seeing beyond its own boundaries and established assumptions.

In practical terms this just means that all concepts are *only* concepts and have less substance than any paper they may be expressed on. They may be extremely useful but can only ever depict 'reality' relatively, not in an 'absolute' sense as they are intrinsically relative in nature, and this applies to the entire range of conceptual thought, however gross or subtle. This doesn't mean that 'reality', before, or beyond, conception, does not exist. Rather, one way of looking at this is that the 'absolute' nature of reality is its ever present immediate *unmeasured* existence, which we tend to constantly overlay with measurement and interpretation, i.e., conception.*[1] The notion of 'absolute' reality being capable of capture in description is simply a contradiction of terms.*[2]

Discrimination and Truth

A central component of the instrument of conceptual thought is discrimination. Discrimination in this context refers to the sense of truth or validity we apply to most information we encounter. We don't do this constantly with everyday things because repeated use tends to reinforce and establish the acceptance of associated concepts at the level of habit. This may relate to straightforward material items, such as the 'chair' in the example above, but it also applies to all kinds of views, attitudes and forms of description acquired through experience or conditioning of one kind or another. Otherwise though, this 'checking for validity' or 'discrimination of truthfulness' is actively directed into all abstract thought to some degree, like a surveyor checking the soundness of building work in progress and is rooted in the established validity of primary concepts, just as the soundness of a building's structure depends on the integrity of its foundations. However, as I have just suggested, habitual use has the effect of strengthening the identification of the conceptual description with 'reality' and therefore of the accepted sense of truth of the description, tending to resist further inquiry in that direction in a form of inertia. This is relevant and necessary for the ways in which we function as organisms. If we had to constantly reconsider every detail of everything we must do, it would be impossible to carry out any complex task, and so we come to accept interpretations based on their functionality, and they become accepted assumptions, a sort of mental equivalent of 'muscle memory'.

The reader will be using discrimination continuously in gauging the validity or otherwise of the ideas being presented here. Discrimination is at the root of logical reasoning, and it's also present in the emotional dimensions of our experience, though not in the form of logic. I will refer to it again, always in the sense of evaluation of truth. We use this faculty all the time, to a greater or lesser degree, in all types of thought, not only conceptual, as an intrinsic element of perception. We apply it at all levels, from fine detail through to that of overall evaluation, it's so basic to our function of awareness we mostly don't appreciate its presence in our attention until it's an overt aspect of some particular thought process. In essence, discrimination is a cognitive faculty that we exercise in both rational and emotional spheres and has firm biological roots in evolutionary terms. Its root is nothing less than the experience of existence, or 'reality', reflected throughout the entire domain of abstract thought as 'truth'. This 'experience of existence', more usually referred to as consciousness, is central to, and inclusive of, all our experience and is the basis of everything that we can consider. The faculty of discrimination underpins and permeates the qualitative evaluations we make in more subtle areas of our experience and social interactions as well as in areas of logic and objective study such as those represented by the experimental and theoretical aspects of science.

These observations don't directly concern evolution at the biological level, but rather, the evolution of abstract thought, language and accordingly, ideas and belief systems - you might say, constructs of information. Nevertheless, principles of evolution - notably, natural selection - that are observable in the organic domains can also be seen at work in the interaction of cultures and collective mentalities expressed through language, where ideas and structures that 'fit' well tend to proliferate and predominate within 'eco-systems' of related and competing ideas. Part of the relevance of this is that what we normally refer to as 'objectivity' is really collective agreement on the accuracy of the conceptual interpretation/description of experiences or events, and this is always relative, never 'absolute'.

Interestingly, there is always some degree of probability, greater or lesser, factored into the evaluation of truth. Without that there could only be ignorance or certainty, no work in progress. Further experience may reinforce or reduce the degree of certainty. 'Knowledge', in this sense, is the concept used to indicate some high degree of certainty about the parts and totality of a body of information, which nevertheless must necessarily remain open and flexible to accommodate new input. In practice distortions can and do arise. Too strong an attachment to a particular form of description, as a belief, makes it doctrine rather than an avenue of discovery and obstructs discrimination rather than assisting it. This tends to happen in all fields of knowledge. Again, from nature's point of view this is sensible. The function of knowledge is practical, and it is acquired through experience, if it supports action effectively it is useful and relied on, not easily dismissed. In a sense, the aspiration to 'pure', or 'absolute' knowledge, is something beyond that, an appetite without boundaries, ever frustrated by attempts to apply it to particulars. You might say that in essence, knowledge is the result of an *appetite for knowledge* with which we are endowed by nature along with the other appetites on which life depends for continuance and development. Ultimately, the significance of all appetites is found in their contribution to the sustenance of evolution.

In practice, all decision making depends on currently available knowledge so there is a tendency for a commitment to be made to the established validity perceived in such knowledge, and to regard it as definitive. This is a naturally occurring and perhaps inevitable compromise of reasoning. Again, this perception lies at the heart of the scientific approach, where as far as possible all theory becomes established only through the convincing evidence of experimentation. In other words, through linking theory as closely as possible with practical experience. A paradox here is that the more thoroughly theory is tied to practical evidence, the greater can be the tendency to convert insight into doctrine.

From these observations so far, one might see that there is a fluid interplay of knowledge between individual and society. However, the individual mind, as the active seat of cognition and discrimination, is the place where the digestion and transformation of knowledge takes place, through what we may call insight. Insight is the ability to recognize deeper, more significant patterns emerging from consideration of prevailing patterns of knowledge, theory, behaviour, experience etc. In important ways though, the individual doesn't exist separately from the society or mental environment they share with others. Our entire conceptual world view from the furniture we use to our more abstract ideas, from the grossest to the most subtle, is shared with the world we inhabit, but with our own unique perspective. But the point at which all changes and transformation, constructive or otherwise, to this conceptual environment originate is in the individual attention. Constructive change comes mainly by the deepening of understanding through insight. This process of transformation of knowledge (of all kinds) informs the collective body of society and contributes to its expanding pool of knowledge.

In everyday conversation, when we speak of 'facts', again, description is confused with 'reality' and an attempt is made to find absolutes in description, which is impossible because description is always relative. Still, collective agreement on the accuracy of a description can afford it the status of 'truth' within a given context, and this is the realm to which 'facts' belong; but description can only ever provide relative truth, regardless of the weight of supporting experience - and evidence - and the further removed the description is from direct experience the more tenuous, or fragile, is that status. To re-employ an earlier example, it is generally straightforward to recognise a 'chair' as a 'chair', and to characterise the use of this term in a factual way that can be expected to withstand the tests of time and opinion, but the same can't necessarily be said about ideas or features of theory that are complex, far removed from direct experience and constantly subject to fresh inquiry. In such fields of inquiry, the notion of 'fact' is used much more carefully, or should be, because fresh discovery and insight may, and generally will, alter the frameworks of understanding and interpretation of experience. These positions represent a spectrum of validity, where most everyday experience of 'truth' and 'fact' takes place between its extremes.

A Little More on Objectivity

In Western culture, inquiry and knowledge has for a long time now been at the service of technological progress. No doubt it is entirely right and proper to use this great ability to improve the quality of life and understanding of its workings, potentially to everyone's advantage, but objective inquiry is a very particular form of exploration, and it greatly influences the way we view the world.

A point I want to make here, briefly, is this: the philosophical basis of the 'classical' scientific view can be summarised as 'objectivity'. This approach is based on and reinforces the 'a priori' perception that everything an observer perceives exists independently of the observer, and that ideally therefore we should aim to construct views or interpretations of the world, which are 'true' and free from distortions introduced by misinterpretation on the part of the individual. This is the basic assumption at the heart of the experimental/empirical approach, which is at the foundation of Western science's great successes. However, on account of the discoveries of relativity and quantum physics we no longer live in the era of classical physics or its philosophical perspective, nevertheless its influence is still great because it remains more generally accessible and useful i.e., more accessible to the intuitive logic of cause and effect. Consequently, we still pay much more attention to the 'objective' dimensions than to the dimensions of experience that shape perception.

Continuing this discussion as I have been, in terms of conceptual constructs, in practice how this works is that we operate a system of collective inquiry and cross reference to create and develop descriptions of 'reality' which are sufficiently validated in this way that we can attribute to them the status of 'objective truth'. However, as I've pointed out, such descriptions are never more than descriptions, and no matter how effective they may be, they always remain entirely abstract and relative mental creations. As such they remain rooted in the field of the 'subjective' whether this is recognised or not. While such an approach creates and develops progressively more complex and powerful abstract models and constructs for interacting practically with the world, its raison d'être and therefore orientation, is that of *prediction and control*. The practical success of this approach means that the tendency to identify with its models of description as 'reality' is correspondingly enhanced and reinforced, on account of the collective weight of conviction and habitual acceptance which becomes attached to them.

A practical observation concerning the nature of objectivity is this; if we want to see how something works, whether out of pure interest, or, more often now, to harness that knowledge, we may take it apart to see what it's made of, what are its components, and what are their components and so on. This approach originates and works well in dealing with tangible circumstances and is also the method we have come to use extensively in the wider spheres of analytical thought. This is fine, but when we do that, we are using the very particular function of our mind that creates 'object' perception, further reducing that object to participating objects, each one an entirely abstract creation of that same mental faculty, conceived of as existing independent of its context.*3

This approach is often referred to as 'top down' inquiry

because it seeks origins through analysis of what can be observed at a gross level, in the present. The complementary approach employed in inquiry is 'bottom up', where attempts are made to recreate the basis of existing circumstances through the (mental and physical) reconstruction of elementary components and principles. However, even in that case, those elementary stages have been conceived of through prior analytical deconstruction, i.e., they are necessarily the result of earlier top-down inquiry, and part of the same system of objective analysis.

We then use the information thus generated to predict, control and construct events. As I've discussed, the tendency to identify the abstraction with 'reality' is strong (perhaps paradoxically, when the individual is aware that a concept or idea is their own creation, identification may be less than when there is collective agreement and support, as that tends to encourage and substantiate the identification).

As I have touched on and now emphasise, the desire to *predict and control* is really at the root of this approach, of 'objective deconstruction', a term I would use to characterise this main way in which the attention of science has been focused for centuries, both in its 'classical' form and still, in its current ones. The relationship between experience and prediction is a very intimate and normal feature of evolution throughout animate nature (considered more closely in Chapter 5), and the influence of this intense disposition on our way of understanding the world, and ourselves, shouldn't be underestimated. It seems to me that at this level the more subtle philosophical ramifications are of little interest within the specialities of science, and rarely have been, because of the focus on control and prediction.

Finally, for the benefit of those who are deeply attached to the idea that reality can be represented accurately through concepts, and are unconvinced by my arguments, I'll round off this chapter with a quotation I particularly like which I've borrowed from 'The Zen Teachings of Huang Po', a translation by John Blofeld of a written record made by a pupil of a Buddhist Zen master in the ninth century ce. To me, this isn't about definition but rather, the opposite - freeing the mind.

'All the Buddhas and all sentient beings are nothing but the One Mind, besides which no-thing exists. This Mind, which is without beginning, is unborn and indestructible. It is not green nor yellow and has neither form nor appearance. It does not belong to the categories of things which exist or do not exist, nor can it be thought of in terms of new or old. It is neither long nor short, big nor small, for it transcends all limits, measure, names, traces, and comparisons. It is that which you see before you - begin to reason about it and you at once fall into error. It is like the boundless void which cannot be fathomed or measured.'

Summary

Clearly, in any observation the position of the observer is crucial. All that is in view reflects that position in one way or another, and in practice, there are usually many factors involved. This is natural and no doubt inevitable and we can't just recognise this in principle to adopt some superior position that permits a perfect and unobstructed view. What we can do though, is to recognise and understand at least some of the factors that contribute to our position and take these into account as far as we can.

These factors largely concern the way in which we use language. At root is the use of concepts as building blocks to construct representations of the world of our experience. Language is an intrinsically collective tool, and accordingly, conceptual representations of all kinds grow through time and circumstance, while generally maintaining some continuity with old historical roots, or fade from use and become redundant, or 'extinct'. These representations considerably influence the way in which we see the world and they are subject to evolutionary process just as everything else is.

Large areas of evolutionary interest are not easily accessible to empirical inquiry on account of the scarcity and fragility of information, and still rest to a great degree on scientific ideas and assumptions that were current in the mid-19th century. The purpose of this chapter has been to draw attention to this and touch on some of the underlying assumptions involved that may condition or predispose our view, before proceeding further.

Chapter Notes

*1   beneath conception lies a deeper layer of perception. This layer is more directly engaged with immediate experience, much more widely represented throughout animate behaviour, and provides the basis from which conception emerges. The conceptual faculty tends to assume a sense of superiority over this level of perception, but it remains ever present in the background as its source, much as the physical skeleton provides the supporting structures of the body.

*2   the term 'infinity' is a good example of the distortion that convenience of use can create in the sphere of conceptual understanding. In current convention, the use of the term 'infinite' derives from a context in which it has come to be used widely to represent the open-ended extension of dimen-

sions of measurement. The root concept however, is 'finite', which essentially means, having a beginning and end in 'time and space', or simply, 'relative'. Basically then, 'infinite' means, 'not finite', i.e., not relative. The nature of conceptual thought, and reasoning however, is to gauge things and events in all-encompassing terms of relationship, and to seek understanding in those terms. Accordingly, in Western thinking, the term 'infinite' has come to mean the extreme extension of measurement rather than its absence. Useful as a mathematical device perhaps, but not 'true', not even relatively true. This observation applies also to the use of the concepts 'absolute' and 'eternal', which, on these terms also reflect the status of undivided reality, before and beyond any relative associations we project and apply.

*3 the concept of the atom itself exemplifies this, originating as a representation of the smallest possible constituent component of 'matter' in ancient Greece, courtesy of the philosopher Democritus around 460 BC. This was long before there was any of the supporting evidence we now have and helps illustrate the essentially abstract mental nature of the concept 'atom'. This isn't to suggest that atoms don't exist, rather, it is just to continue the point that description and actuality are distinct and separate. On this point, it should also be borne in mind that although atoms are the smallest basic particles as far as the nature of elements and their interaction is concerned, they are no longer regarded as being the smallest particles of matter, being composed of even smaller components essential to their structure which can scarcely be even conceived of as 'particles' in a conventional sense.

Chapter 8

The Domain of Culture

Outline

The central topic of this chapter considers the social behaviour of the human species as reflecting the same basic principles of evolution that can be observed throughout organic life generally. Following on from the discussion of consciousness and intelligence, it is clear that conscious interaction between organisms has become an important factor in the directions that all subsequent developments have taken. This is a medium in which co-operation and competition take place, just as much as in more immediately physical spheres of engagement, and ultimately have physical consequences at all levels of developments. Accordingly, it is a sphere in which all the faculties of experience continue to develop and refine. In my use of the term here, 'culture' refers to the entire range of human social engagement, as an evolutionary domain. In this chapter I look at some of the ways in which this is represented. This isn't intended to be a definitive or detailed inquiry into the workings of culture; that would require a lot more than a few pages and is far beyond the range of my knowledge. It's just an outline that attempts to put the idea of culture into an evolutionary context as an important element in the overall discussion. There are patterns of relationship here that reflect essentially the same kind of principles that can be observed throughout organic nature, particularly natural selection through competition and co-operation, and the ongoing development of complexity supported by equilibrium. There can be little doubt that viewing human social behaviour and culture through the lens of evolutionary insight can contribute to our understanding of events, past, present, and future.

Later in this chapter I discuss the role of values in the developments of culture. Values in human society can be understood as rules of nature as they pertain to our sphere of experience, and therefore reflect evolutionary principles in action. This isn't a new observation, the notion of 'natural law' underlying all living behaviour has deep roots in Eastern philosophy, where it is referred to as 'Dharma'.

Normality!

Normality, it's so simple and uncomplicated, mostly we just take it for granted. The world we are used to and live in from day to day, with its comforts and familiarity, its major and minor hassles and tensions, its occasional unexpected events, and dramas. You look, you see, you experience the world around you. Here is reality! Obvious, straightforward, right there in front of you... But as with many things, when you start to look closely, there is much more going on than first meets the eye...

The 'normality' of personal experience is intensively structured, or shaped, by the environment we inhabit. This shaping proceeds from the moment we are born, and in some respects, even earlier. The environment of course comprises the physical and mental features and circumstances that surround us, but the 'shaping' also reflects innate qualities of our own responding to those external circumstances - as ever, the combination of 'nature and nurture' is an immediate and all-pervasive condition. From the very beginning, we absorb information about the environment - and impressions of ourselves as reflected in those perceptions. Centrally, this is direct sensory and emotional experience, but combined with this, and no less commanding, is a world of abstract description communicated and acquired through language. This influence is more subtle perhaps but is a sphere in which the human social environment is now largely shaped, so it's by no means superficial. Where sensory and emotional experience reflect our inner genetic heritage directly, abstract description represents the ever flexible 'software' that shapes that experience. As with all the practical things we learn throughout life, we absorb these influences quickly, especially when we are very young, and move on. Details become stored; like muscle memory, we don't have to keep relearning the same things again and again or we would never be able to move on to other things and so, our personal sense of 'normality' is, to a large extent, moulded over time through the cumulative establishment of patterns of repeated experience and behaviour. This shaping applies to both the emotional and rational spheres of our experience. On one hand then, innate qualities of the individual are at the heart of their experience, on the other hand the dimensions of this experience are influenced directly and in an ongoing, cumulative way by the environment. As in all evolutionary contexts, this relationship is reciprocal and is at the core of the domain of culture; the environment influences the individual, the individual influences their environment.

The principal instrument of personal experience and collective engagement involved here is, of course, the brain, which doesn't appear to have altered much physically in at least the last 200,000 years. Clearly though, the circumstances that we learn to recognise as 'normal' and which generate the way in which we see ourselves, are very different from the kind of circumstances that one may have experienced in such ancient times, particularly before the development of agriculture and cities. So, if the principal changes in circumstance aren't a result of biological changes within the human brain itself, what is the medium of change? The answer, simply, is the sphere of *collective interaction and shared behaviour*. This is what we refer to as 'culture', and its central features can be outlined broadly as learning, communication, and sharing of knowledge - all of which have roots in memory. These factors provide the basis for the ongoing development of complexity beyond the strict limitations of the genetic or molecular spheres (though these spheres of course remain integral). As far as these features are concerned, we are by no means unique. Many other species express them also, but the degree to which ours are developed does appear to be uniquely extensive, with important consequences for the direction evolution is now taking overall. In effect, human culture is now the dominant domain in the evolutionary hierarchy of this planet.

Because the concept of culture is intrinsically one that concerns collective behaviour however, when we think about 'society' or 'culture', attention is naturally drawn to focus on events in terms of patterns of collective behaviour, and to underappreciate their roots in the individual psyche. For many practical purposes of social organisation this may be unavoidable, but for reasons that I'll discuss, it seems to me important not to lose sight of the principle that the active core of collective behaviour lies at the level of individual perception and behaviour. This relationship can be compared to that whereby the variety of element and molecular organisation rests on the innate structure of the atom. The key to understanding molecular processes lies in understanding of the structure of the atom; similarly, human social behaviour and organisation reflects the make-up of the psyche. Individual people interacting with each other *is* what a society is, there is nothing else. Accordingly, great complexity arises, and social organisation progresses in ways that reflect evolutionary principles represented in all domains. The principle of hierarchy is significant here, but this isn't represented in ways likely to conform to expectations we may conceive of in terms of man-made structures of organisation. Many patterns in culture can be identified by observing past events and such patterns may be recognised as they recur in different settings, but the forms that natural complexity creates are never entirely predictable, with often major and entirely novel shifts taking place in collective attitudes and behaviour. Nevertheless, behaviour at the greater level always reflects, in a variety of ways, the nature of the contributors at the underlying level. In the case of society/culture, this is the make-up of the individual human psyche.

As I've discussed at some length, evolution as a phenomenon isn't restricted to the physical dimensions of circumstance, though these always participate, but more fundamentally, concerns the development of complex organisation rather than of physical particulars. As such, the same basic evolutionary principles and patterns of expression that can be observed throughout nature, inanimate and organic, that we have been considering, are also represented in the domain of culture. To help to put this into context I'll briefly recap a little. Properties of behaviour within any particular sphere are largely conceived, and described in terms of, principles which are derived through the observation of the phenomena of that sphere, e.g., at the level of molecular studies the terms are those the physics of the atom and molecular chemistry; in the context of organic life they are conceived of in terms of cellular metabolism, genetics, biochemistry, survival, reproduction etc; in the context of collective human behaviour all of the principles of these domains still actively underlie and participate, but generally don't suffice to provide descriptions of the way in which developments proceed here. Patterns of social behaviour tend to be described more typically in the language of psychology, anthropology, or sociology, which attempt to apply similar methods of empirical inquiry to those employed in the physical sciences, but with much less available in the way of tangible/material evidence. What can often be under-appreciated in the inquiries of these various fields though, is the evolutionary background to behaviours. While the details of inquiry and description, are always relevant, what interests me more is the expression of evolutionary principles and patterns common to and underlying all levels, and how they are represented in the field of collective human experience, such as:

- the relationship between the individual person, with their intrinsic qualities, and the social environment, with its complex, variable, and fluid influences; the 'ecology' of social context, you could say.
- the ever-present tendency of natural selection to support equilibrium and reproducibility, over instability and entropy.
- the balance between co-operation and competition as a central feature of self-organisation.
- the nature of freedom of individual behaviour within the limitations of circumstance. (Not inquired into in detail in these discussions.)
- the interpretation of motivating forces in terms that relate to the context of the domain. In the case of 'human culture', this wouldn't be those of the atomic or molecular spheres, or biochemistry, i.e., energy as conventionally conceived of in dedicated physical terms, but 'social forces' or dynamics. As with those other forms of 'force', these need to be character-

ised in terms derived from the observation of dynamics of the domain in question, culture, and rooted in the qualities of the participants - which in this case means innate qualities of the human psyche. I'll be exploring this matter a little further in the following pages and chapters.

- the likely emergence of new forms of expression of consciousness within the domain of culture, since it, consciousness, is at the forefront of evolutionary developments there.

While the overt characteristics of all domains, from those at the molecular level to those of human culture, are very different, the underlying principles of their evolution remain essentially the same, and continuous. In considering developments within the field of culture it is worth bearing this in mind.

Cooperation and Competition in Culture

Co-operation and competition comparable to that of organic nature, are prominent features of evolution in culture that can be seen represented within and between human societies. The results of this in physical terms are reasonably plain to see and are observed and surveyed through studies such as history, anthropology, and archaeology. Equally important to the more dramatic events of historical observation, is the underlying exchange of information and ideas. This exchange is in effect co-operative, regardless of intentions or boundaries of ethnicity, nationality, politics, or religion. Physically, civilisations and empires come and go, typically as a result of competition in the form of military conquest, but with much more intimate roots in the ongoing development of ideas and knowledge that permeate across all barriers and are integral to the emergence of complex civilisation.

As is well recognised, developments in agriculture were significant in the emergence of large scale organised human societies. But the sphere of agriculture represents only one of many areas where the sharing of knowledge is an essential factor. Practical inventions and discoveries that prove useful tend to stick around, regardless of who comes to be in charge at any point in time. All such innovations are creations of the mind, and continue in all sorts of fields of inquiry, from the invention of flint tools to those of mathematics, from control of fire to information technology. (Of course, this 'co-operation of knowledge' applies equally to the developments of warfare technology, despite all efforts at secrecy. Warfare remains a significant factor in the competitive side of cultural engagement - and invention - but economic pressure has frequently been a significant factor in the occurrence of war and other forms of social change, and among larger modern societies physical warfare is increasingly being replaced by competition in the form of economic pressure.)

Where agriculture is concerned, invention and development continued into related spheres, such as trading networks, and systems of accounting essential to large scale management of materials; at the same time, the most useful discoveries and inventions largely survive the demise of the societies that create them and are adopted elsewhere, simply because they are useful generally. They may be adapted in different ways in different societies, but they belong to no-one and everyone. Core creations range from the most basic and practical such as again, control of fire, or the invention of the wheel, to the more subtle; ideas, concepts and (later) writing; also, insights of religion, philosophy, and the sciences as well as social ideologies such as democracy, communism and (effectively) capitalism. As ever, the balance of co-operation and competition supports the preservation and development of those mental creations that contribute to successful social organisation, and the ultimate demise of those that don't. At the same time, this balance finds ongoing expression in the present-day dynamics of culture, as philosophic, religious, political, and scientific ideas, beliefs, and ideals 'compete' for position in their own areas of influence. Principles of equilibrium govern and sustain this. Just as 'nature' and 'nurture' work together in shaping form in all living domains, co-operation and competition function together and are inseparable partners. I've discussed this relationship in Chapter 4, so I won't repeat it here. Suffice it to say that 'competition' effectively means 'competition for resources', and 'resources' range from the basic requirements for the chemistry of cell metabolism to the information and ideas requirements of human civilisation. This principle continues to apply throughout the abstract dimensions and creations of group human culture and collective behaviour. At this level, of human culture, the idea of resources necessarily still includes basic physical requirements, all the way down to the molecular level, but expands into much more complex and abstract areas of common interest. It's worth commenting here also that co-operation is by no means necessarily 'voluntary'... From the domain of molecular organisation to that of human culture, the pressures of circumstance, gross or subtle, frame the context and options available.

As in the case of molecular organisation, with the 'sharing' of electrons, in the context of culture the idea of 'sharing', is central to that of co-operation. The simple act of observing the behaviour and ideas of others means that influences are spontaneously shared and absorbed through learning. We share and exchange views, ideas, insights, theories, opinions, analyses, feelings, and values generally, primarily via the medium of language. 'Sharing' is the sea we swim in together and is the basis for the continuity and growing complexity of all our mental creations. 'Sharing' in this sense is at the heart of culture and is the expression of co-operation in a broad evolutionary sense, within this domain. I've discussed the general basis of this view of co-operation as a principle in the section on natural selection, and as the discussion proceeds, I may offer some more comments on the relationship between co-operation and competition. A basic one though is that whether we view situations in terms of competition or co-operation largely depends on the stance we take as observers. It can be said that 'competition' reflects 'co-operation' taking place in a broader context (the opposite is also true). In a sense, the entire field of ecology as a self-balancing phenomenon throughout all domains, can be viewed with this in mind.

Perhaps the most basic expression of co-operation in social organisation can be found in family relationships. A balance between co-operation and competition exists here just as it does elsewhere, but generally, the 'centre of gravity' appears firmly within the frame of co-operation. Of course, competition exists as a part of family relationships in many ways, but within the boundaries of family the dominant theme is generally that of people living, learning, and sharing together - essentially a nurturing environment, with more distinct competition for resources largely emerging between different family groupings, but with still a broad sense of that ethos of co-operation reaching out through the extended family, in forms we might think of as tribal. At larger scales of social organisation, this inherently cooperative sense of 'family/tribal bonding and identification' is found recreated in many spheres, trivial and more significant, and far from its roots, from football club support to nationality, race, religious and political affiliations, with a corresponding tendency to competition with other such groupings. Beyond its origins in actual close family, in practice this pattern of 'family' identification may require only a small degree of common interest or belief and can be seen to be highly flexible. 'Brotherhood' and 'sisterhood' are common themes of solidarity, as is the tendency for leadership within societies to follow patriarchal or matriarchal patterns of authority and governance.

However, a broader sense of the common interests and rights of people of all societies has been growing over recent centuries - along with an understanding of the inherent unity of life - resulting in the perception of humanity as one 'family', transcending boundaries of nationality, ethnicity, and religious belief. This can be seen as entirely idealistic, contrasting as it does with a long history of empires and illusions of racial, cultural and religious superiority, but seems to me to reflect the central role that the sense of family ultimately plays in the dynamics of social organisation at all scales. It becomes apparent that the interactions of group identities, at all levels, are subject in their own way to evolutionary forces. This outline can give only a very basic sense of developments rising through the relationship between co-operation and competition in the cultural sphere. Still, it seems reasonable to consider that the balance of these two aspects of behaviour dynamics is a feature of the developments of coherent complex social organisation, and the shaping of cultural form at all levels.

Values and Sustenance

Commonly, we may observe evidence of competition or co-operation within and between societies at the level of overt forms of behaviour and interaction, but when we think of culture as an evolutionary domain and look for patterns in behaviour that bear signs of expressing evolutionary principles, a more subtle but all-pervasive sphere worth observing is that of *values*. Clearly the coherence of social organisation comes through co-operation in one form or another. As elsewhere in nature, refinement comes through competition - what works best persists and can further develop. A result of all this is cumulative experience, through which certain impressions persist in a residual form in our collective attention as 'values', acting as protocols, or guides, to behaviour in the areas of both co-operation and competition. The term 'values' in this sense isn't limited to the sphere of ethics or philosophy we may generally associate with the word, but also relates to most practical areas of everyday social interaction. In a still broader sense 'values' are reflected in the aspirations and ideals active in society, as well as acting as guides in the direction of behaviour.

Values

Although we usually talk of 'values', in the sense of nouns, it would probably be as well to remember that they originate through 'evaluation', which is an *act of judgement of quality* that takes place in the first instance in the individual mind. In common use, the term 'values' usually takes on the more general sense of social 'mores', representing collective consensus concerning views and attitudes. We can see from this that there are two sides, or aspects to the meaning of 'values'; the individual contributes their personal faculty of discrimination to *evaluate*, while the social environment tends to absorb and dominate individual contributions. Over time, shifts in collective attitudes take place and new patterns, or variations, of mores emerge. The underlying relationship involved is, as ever throughout nature, that of the contributing element, or component, with the environment.

One way of understanding the function of values is to compare them to the way in which we use road markings, which is to indicate collectively recognised rules, or guides for behaviour that make things easier for everyone to live to some degree in harmony with others. For example, lane markings make it much easier to anticipate the likely movements of other drivers, speed limits reflect local safety conditions, traffic lights generally help to avoid congestion and so on. Although all these methods represent elements of restriction on the freedom of movement of individual drivers, in the collective situation they make decision making and freedom of movement easier for everyone by establishing agreed protocols, or 'rules', of behaviour. Similarly, for all kinds of society, over time and experience, rules of behaviour are formed, and for the most part, their function is to facilitate harmonious relations and freedom of movement throughout many different aspects of social intercourse. For that reason, although there may be a tendency to perceive or discuss 'values' in static terms, really their role very much concerns their guiding influence on the *dynamics* of social behaviour - the way we relate and behave towards each other. In this sense, values are intrinsically principles of co-operation.

Extending this analogy a little, clearly the regulation of traffic in a city is much more demanding and complex than it needs to be in a small rural village. Nevertheless, regardless of the size or complexity of the situation, certain basic common rules must prevail so that people can navigate throughout without encountering conflicting rules of behaviour. In larger conurbations, the 'language' of such signs is essentially the same as in smaller ones, but the 'vocabulary' becomes more elaborate, and more specialised guidance may be required for the navigation of particular areas. There is an element of similarity with values in the sense that the more complex the society is, the more variety of spheres of interest there are, and pressure for defined 'rules', and 'protocols of behaviour' to adjust to local circumstances. Of course, the sphere of values is very much more diverse than that of traffic control, with many spheres of concern, from most basic practical issues around survival to matters aesthetic or spiritual. Systems of law attempt to apply collectively agreed principles of value in the concrete form of legislation, i.e., enforceable rules - laws. At its best this can contribute significantly towards advances in social justice, but it's also a means of controlling and manipulating social behaviour to very different ends. When construct rules come to outweigh and overrule personal discrimination, this primary faculty - which after all is at the core of ethical behaviour, value creation and social bonding - can become undervalued and suppressed, with profound social consequences.

All this range of concerns reflects and bears influence on the individual even when they are largely unaware of it. Each person has of course concerns particular to their own situation, from those of everyday practical matters to their more subtle aesthetic aspirations and engagement. The individual psyche is at the heart of all social organisation but clearly no one person, within their limited circumstances, experiences or represents all the dimensions of concern to the collective well-being. However, at the collective level a very broad range of interest is represented, and although much of that may be of little direct concern to any given individual at any point in time, still, unvoiced aspects relevant to their existence are being addressed somewhere and contribute to the collective 'pool' of values.

From an evolutionary perspective, this wider, collective, sphere of representation is where 'values' have their greatest relevance, as it is the context within which both the collective psyche of humanity as a species, and consequently the individual psyche, continue to be shaped and evolve. It is here that the greater potential of the psyche necessarily works out. Within the collective domain, through the decisions we make - and act upon - we select which aspects of the human psyche we restrain and which we encourage, i.e., cultivate. In evolutionary terms, you could say that this is the function of culture as a domain. At the level of biology, we are provided with a broad base of instinctive motivation and impulse inherited through our deep and ancient ancestry. We can't just override or change that, or pretend that it doesn't exist, but through our collective engagement we continuously select and shape human behaviour, restrain or encourage the desirability of particular features. This is the realm of values, at the heart of culture and the ongoing development of the psyche.

It may be worth observing here that an implicit recognition of this principle of cultivation resides in and underlies the entire field of values, i.e., of preferring certain attitudes and behaviours over others. Accordingly, ideologies develop in one form or another, religious, philosophical, or political, but ultimately, evolutionary forces and principles of selection govern the directions and outcomes of events rather than any particular views, interpretations or ideologies.

I'll develop a little further here an analogy I used earlier that may be helpful here for some readers based on common, modern, experience, which is that of computer hardware and different levels of software. The 'hardware' aspect of a computer system is the physical basis of its overall capability. As far as 'software' is concerned, this relates to the information handling capability of the system. In the first instance, this must be compatible with the hardware capability, and is known as the 'operating system'. The purpose of this level of software is to integrate and coordinate the hardware with the widest range of possible uses that it may encounter. The next level of software is that of 'application'. As the word suggests, application concerns the actual use to which the hardware and operating system will be subjected. This is a much more flexible/adjustable sphere of software function, but again, it has to be compatible with the operating system for it to work. Comparing this to the human situation, the hardware takes the form of the brain and nervous system. In our case, unlike in most modern computers, the 'operating system', which comprises the innate structures of the psyche, is provided by nature along with the 'hardware' because the two aspects evolved in conjunction - but is nevertheless still, in effect, 'software'. This 'operating system' is subject to evolutionary developments on the normal time scale of genetic processes and has been established in much the form we express now as Homo Sapiens, probably for around 200,000 years. The 'application' level however is presented in a constantly flexible form by the prevailing environment through the medium of culture. The time scale of significant developments in this domain, conscious culture, is very much shorter and constantly shrinking, from thousands of years, through hundreds, now perhaps across a few decades.

One point of this analogy is that while the genetically inherited 'hardware' and 'operating system software' provide the broad scope of potential and limitation for the capability of our organism, this can only ever reflect the options actually presented at the level of the 'application software'. In the case of a computer, regardless how powerful or well designed the hardware and operating system are, the results are always shaped and determined by the 'application' information presented. (If you program a supercomputer to play 'pong' and nothing else, that's all it will, or can, do.) In the case of human beings, the 'application software' takes the form of the cultural context as it is at any point in time and space, rapidly and continuously encountered and 'downloaded' from the moment of birth.

In these terms we might say that the faculty of discrimination, as an innate feature of perception, is seated within the deeper level of the 'operating system', while particular values, as 'rules of behaviour', result from the interaction of that feature with the conditions and demands of the prevailing environment (the 'application' level). It's worth considering though, that value discrimination primarily concerns morality and the emotional dimensions of experience, not rationality, though rationality could be said to participate in a supporting role. ('Truth' is a central quality of evaluation through discrimination of all forms of experience, including rationality, but rationality alone, as the mental expression of logic, is inherently devoid of moral dimensions.) All the qualities within discrimination are, naturally, deeply rooted in our evolutionary heritage. These ancient roots bear an ongoing living influence within the living sphere of values.

A profound result of evolution within the domain of culture is the shaping of human mentality beyond the strict definitions of biology, it's refinement you could say - with broad material consequences as well as mental. Values are guides we create in that process. However, in the analogy with traffic control, road signs are no more than that, they are inert and exist externally to us, but do values have any deeper significance for us beyond that of simply advising the flow of traffic...? In exploring this question another, more organic, comparison that I've found useful in understanding the place of values in human culture is the structure of a tree.

In a tree, nutrition is transported in the form of sap from the roots via the trunk and the branches to the extremities, where the leaves, under the action of light, transform the raw material to new growth. Strength is an integral feature throughout the structure, contributed to by the roots, the trunk, and the branches. That strength is represented variously in forms of both rigidity and flexibility. Accordingly, at the same time as carrying nutrients to and from the leaves, the branches provide structural strength to all they sustain, and distribute and balance the overall load of the tree. Interestingly for the purposes of my analogy, the tissue that carries the sap from roots to leaves, called xylem, is largely composed of dead fibres of many past years, partially decayed to form passive capillaries. These work in conjunction with living tissue fibres called phloem which actively distribute nutrients generated at the leaves throughout the entire organism. Most of these processes take place in an outer region of the trunk known as 'sapwood'. The core of the trunk is composed of fibres which are further collapsed and no longer act as capillaries but continue to provide much of its structural strength and resilience. Overall, water rises from the roots via the xylem, drawn by transpiration (evaporation at the leaves in combination with capillary action), carrying nutrients to the leaves. These nutrients are chemically transformed by sunlight at the level of the leaves and redistributed to support new growth throughout the tree. Structural support and the transportation of nutrients is therefore provided by a combination of the passive contributions of its past growth and the active creations of living tissue. In this way the tree's vigour is sustained. Many of the fibres of the tree's structure are no longer strictly speaking 'alive' but are nevertheless essential, and integral to the sustenance of the organism. The reader may be relieved to know that I won't attempt to explore tree anatomy any further here as it's not necessary for the purposes of the analogy, but I do particularly like the image of the way in which strength and growth in the present is sustained by that of the past; even though that past is no longer active, aspects of it make significant contributions to sustenance and developments in the present.

This image is intended to help illustrate, that like the trunk and branches of a tree, values are much more than simply inert abstract concepts. Having origins in discrimination, their roots are deep in more ancient regions of our cognition directly connected to our biological heritage, and are intrinsic to the structures of collective behaviour, social relationship - and to self-identity, individual and collective. Within the abstract sphere they develop and adapt over time, providing strength and sustenance to societies. Even when their origins are long faded and forgotten 'values' continue to draw nourishment from roots in the unconscious (to venture momentarily into psychological terminology...). We could say then that value discrimination originates at a deep instinctive level within the individual but is shaped and modified by collective experience in the abstract domain of culture - the environment. The manifest 'values' this creates act, as-it-were, as supporting branches in the 'tree' of culture, contributing to its sustenance. For the most part these are unconsciously present in our collective psyche, but they become consciously expressed at the level of the 'leaves' (to continue the analogy a little), i.e., in the everyday interaction of individuals with their community. This is a vital area in the life of our 'tree', where essential ongoing new growth, transaction and transformation of values takes place, and where older forms are refreshed and reinforced or re-interpreted.

One final observation I'll draw from this analogy is that all the functions of the tree; the trunk, roots, branches, leaves and all its processes of growth and development are integral features of its existence as an organism. No part is 'more tree' than the others. Similarly, the values that we create through the exercise of discrimination are not just instructions for guidance, but reflect our nature as an organism, moulding and refining over time and circumstance. Throughout all world cultures, the relationship between individual discrimination and social 'values' is a significant factor in the shaping of a society and the directions it moves in through time as a balance between them is constantly sought. For this reason, again, the field of 'values' is a good one for observing and considering cultural evolution, because it reflects the central principle of equilibrium, underlying the development of the psyche, personal and collective.

Value judgement permeates every aspect of social existence, from the most trivial to the most important, the simplest to the most complex, from the level of mundane everyday matters, practical and abstract, to those of justice, legislation, global politics, and religion. Also, mostly when we think of values we tend to think of morality and ethics, but the discriminations we make and find extending into the dimensions of cultural form include the appreciation of aesthetic qualities such as beauty, harmony, and all forms of artistic and dramatic expression, as also, the recognition of 'truth' in the sphere of logic and reason. In some respects, aesthetic discrimination more clearly exhibits the reality that deeply rooted, innate, sensitivities underlie the forms and shaping of culture, not 'ideas' alone, and are omnipresent features of value judgement.

Values are by no means one dimensional, they are represented throughout all forms and levels of cultural organisation. They are collective expressions of the sensitivities and perceptions of the individual psyche, which has many dimensions. But let's limit ourselves to conceiving of culture, and the values that sustain it, as a 'three dimensional' world of everyday experience. Clearly some values are held to be more important than others, and this is reflected in the background of all the cultures and civilisations that we have some knowledge of historically, and in many of these there appears to be a perception of man's hierarchical positioning within the universe, however conceived. Value systems and prevailing structures of social organisation seem to have always reflected this in one way or another. Values don't only serve the purpose of sustaining the stability of society in a passive sense, but like the branches in our metaphorical tree, they also reflect its aspirations towards further growth and development, and consequently, the greater potential of its nature. In this sense values form conscious and unconscious guiding dimensions for our further growth, individually and as societies. Effectively, they reflect and represent an 'ecosystem' of social intercourse in the domain of conscious culture, which in many ways parallels that of the organic sphere.

As a society grows and develops many values of longer standing remain, some are further refined, such as the right to life and to quality of life; some may become less clearly defined, for example, in these times, those concerning sexual mores and marriage; priorities shift according to demands and pressures in the collective attention. Also, there is effectively, 'competition' between and within values of different times and cultures, or aspects of them, as and when they meet each other - which is increasingly the case - and sometimes conflict. This competition takes place in different ways and on different levels, from that of direct social engagement, to that of philosophic or religious ideals. In one way or another, all of these contribute to the balance of social chemistry through which collective value systems evolve.

In the final analysis it's not so much human opinion or ideology that prevails, rather, natural selection within the field of values exerts its judgment in the shape of stability in the longer term, including the persistence of particular forms - which is not always obvious in the short term. For the purposes of this exploration my focus of attention is on the general principle of equilibrium, as a central feature of coherent development throughout all domains of complexity. I would say that this isn't just an exercise in idealism or positive thinking, but an observation of the workings of nature, as reflected in the sphere of human behaviour.

For that reason, I don't go much into the negative aspects of value developments - which certainly exist - in this short discussion, but it should be considered that many of the same principles that are expressed in the development of constructive values are also represented in that of more destructive forms. Those are just less likely to stick around in the longer term owing to their inherent instability. As ever, natural selection is the arbiter of what is constructive and what isn't, what sustains and what doesn't, not our personal views, hopes, fears or expectations. There is much more worth considering on this topic, but perhaps not in this discussion.

Values and Identity

Values then, emerge and take shape through the dynamic of individual and collective relations, and they contribute significantly to the nature of group identity. The individual attention is intimately subject to the influences present in the environment, which initially and to a large extent ongoing, take the form of family, immediate and extended, and soon, peers within the local and broader community. This development proceeds at a largely instinctive level of emotional exchange, but from a very early age, personal discrimination is a significant element which, in conjunction with the environment, is where 'identity', both personal and collective, becomes more defined. Self-awareness also, is an intimate factor throughout all of this and is integral to both individual identity and culture as a collective phenomenon.

Naturally, the individual psyche fits in well with this for the most part; after all, the entire nature of cultural evolution reflects and is part and parcel with, the human psyche, so they fit like hand in glove. The active evolutionary drive is seated in the individual - which is where all live transaction takes place - and is reflected in the dynamics of collective behaviour. A large element of the form this behaviour takes reflects tension between the inner discrimination of the individual and the values/mores currently represented in their society. This in turn is reflected in the search for coherent identity on their part, and a corresponding continuous discourse around values and identity within society.

This discourse is so pervasive that it may not be immediately obvious, but in virtually every news article, movie, soap opera, courtroom consideration or even bar-room conversation, value considerations are expressed and exchanged, directly or implicitly. On looking around it is clear that we swim in a sea of value judgement of one kind or another. This is an entirely reasonable position to be in, reflecting pressures active within the domain of culture, in which identity and identification play a large part. Depending on the circumstances, these pressures are likely to range from social factors of an economic/political kind, or environmental issues, such as atmospheric pollution and climate change, through to the direct and indirect consequences of religious conditioning, which has a long and ambiguous history. Commonly, attention tends to focus on such areas, leading to action, such as protest, political organisation, new legislation, revolution, war, and civil war at the extreme. More subtle internal reflections on identity, may appear to be drowned out in the drama of such pressures, but they also are an important expression of the aspiration towards self-esteem and dignity that is also an essential element of social dynamics.

A common reaction to social pressure is to shrink from it

in one way or another, but once discrimination is stirred out of relative dormancy, it doesn't easily retreat into its shell - and it is contagious. Accordingly, many people become motivated to seek a more substantial source of values and identity than may be available by convention alone. Under all such pressures, great or small, one consequence is that recognition grows, within perhaps a small percentage of the population, that ultimately moral decision resides internally, within the attention of the individual, rather than in the 'traffic signs' of convention in the shape of doctrines of social definition, religious, political, or legal, as these become increasingly difficult to interpret and apply. It seems to me that this recognition is always emergent to some degree as an inherent part of human nature and in the balance of cultural progress, as, after all, personal discrimination has always been an integral element of social chemistry. At certain points, individuals have value insights arising from significant pressures of the time in their society which they observe, that may conflict with, or simply diverge from, established norms, conventions, and rules. These insights may then become part of the collective value system if they are sufficiently resonant. This is apparent as a factor in the emergence of all the world's religions, philosophies, and scientific discoveries. Although that's reasonably self-evident, what is perhaps less obvious is that this factor is also ever present at the level of everyday personal interactions, and this is reflected in the appreciation of cultural 'heroes' and iconic figures in all fields of society, from medicine to politics, from scientific discovery, to sport and the arts.

This shift towards internal inquiry is only one consequence of modern-day complexities, other perhaps more obvious ones are; social deprivation, identity confusion, insecurity, vulnerability to manipulation, alienation, and associated degradation through drug and alcohol abuse, to name but a few. A factor common to all of these is loss of self-esteem, which I'll discuss further. From an evolutionary perspective, these effects are symptomatic of unstable community, within which individuals become emotionally stranded, while the search for a significant value base reflects the pressure towards stability. There is also a reaction to this growing tendency towards personal discrimination that takes a variety of different forms. Perhaps the main one of these is the fact that any established system of organisation naturally involves a substantial degree of inertia and social acquiescence. 'Inertia' is an implicit element of stability in any context. Socially, a certain degree of personal freedom is essential and enjoyed, but not any degree that might threaten the status quo of established order. Individuality and freedom of thought always represents a potential threat to such order, and reaction takes shape in the attempt to maintain control by whatever collective body feels pressured, at any scale, and in proportion to the degree of threat perceived. This occurs in all social bodies, from the level of families through to that of religious and political organisations, institutions of all kinds and all shades of collective organisation in between.

Again, as natural selection always supports stability there is reason for optimism here in the long term, since the evidence of evolutionary history is that elements promoting and sustaining group cohesion, i.e., co-operation, generate strength and are as a result selected, naturally. Inertia is an essential feature of that, but by no means, on its own. Stability requires flexibility as well as rigidity and evolution is inexorably dynamic. Where inertia restrains too much, it no longer contributes to growth and is bound to give way sooner or later through decay and collapse.

From this perspective, the aspiration towards meaningful and supportive values is at the forefront of the evolution of culture, and the root of this process is actively represented in the individual's search for meaning, identity and self-esteem. Viewed in these terms this is not just a field of random or incidental events but is an important aspect of evolutionary process whose directions and outcomes are likely to be of great consequence to the future of human society and ultimately, what it means to be a human being. Still, it should be borne in mind that evolution is 'long-sighted' and entirely impartial. No particular form is indispensable...

This stage in the developments of culture, Western at least, is tending therefore to produce considerable changes in the way values are understood, with a shift in balance towards recognition of the importance of individual discrimination over established conventions and mores per se. Rather than relying on rules and norms the individual is under pressure to exercise and trust their own judgment more. In the short term this presents challenges for everyone, but in the longer term it may well contribute towards the development of stronger collective decision-making and creativity, and to the cultivation of refreshed value perception. That doesn't necessarily mean the loss of all older values. Many qualities of value perception are likely to be deeply established within our 'operating system', not necessarily in definable form, but rather, as elements of our field of instinctive emotional response. Over time, often great periods of time, expression of these instinctive responses becomes refined through cultural conditioning, but never removed - being rooted at the level of our genetic heritage, and there are no doubt layers to this refinement also. In some instances, these ancient and ever-present roots may come to be re-cognised and better appreciated. In the case of some of the more newly prominent values, such as aspirations towards human rights and ethnic equality for example, it seems reasonable to suppose that these don't represent entirely new aspirations, but that current times have generated the collective conditions for their emergence into wider recognition. In general, one might observe that wherever injustice is experienced in society, locally or globally, social evolutionary pressures become active towards establishing balance, though there is usually a great deal of inertia and resistance to change involved. In any given case, change may appear glacially slow, and by no means inevitable, but like a glacier, evolution through selection demonstrates irresistible force which can be relied on to bring results, though not necessarily what we might attempt to predict...

Summary

It seems clear to me that evolution represents a process of coherent developments in complexity arising from principles of nature expressed throughout every level and domain. This includes the domain of human culture. Natural selection plays its part in the developments of culture just as it does elsewhere. This can be understood in terms of co-operation and competition which can be perceived in all contexts. While this does have overt physical consequences, at a more subtle level it is reflected throughout the entire field of values that underlies and influences behaviour. This 'field of values' is at the heart of cultural evolution, and its consequence is the ongoing shaping of the human psyche.

Chapter 9

Identity and Self-esteem

"Oh, wad some power the giftie gie us,
to see ourselves as others see us."
(Robert Burns)

Outline

The last chapter focussed on human culture, and there were really two main reasons for that. In terms of the hierarchy of evolutionary developments, it's clear that human behaviour, collectively, has reached a stage at which it is having a major influence on the directions evolution is taking on Earth. We are no longer passive participants in nature, and looking back, it's clear that we never really have been, it's just more obvious now that we are confronted by some of the negative consequences to the planet of our behaviour to our environment. Accordingly, we are faced with the challenge of understanding and facing up to our position within nature, not just in terms philosophical, but in the fullest practical terms that we can muster. This means understanding how we think and act collectively.

The other reason is the same one that has lain behind all kinds of human inquiry for time immemorial of course; simply, we want to know about our own existence, how we relate to the rest of life and the universe of existence. This may appear to be a purely philosophical fascination, with little practical relevance, but my own view is that these considerations are really two sides of one coin and should be addressed together. The philosophical side is an intrinsic and relevant feature of human mentality, a major aspect of social evolution that can't and shouldn't be sidestepped, but many aspects of human behaviour make much more sense when their evolutionary background is taken into account. The area that this chapter and the remaining ones consider, concerns the very intimate relationship between the individual and their environment, which necessarily involves their psychological/emotional well-being. This, in my view, is where at least some insight into the directions of evolution lies.

In the last chapter considerations mainly concerned the general idea of culture as a domain in which evolutionary patterns can be observed. This chapter looks a little more closely at the principal component that underpins the dynamics of culture, namely the human psyche.

Cultural developments are generally viewed through filters of anthropology and history, but these methods of study observe events primarily at the level of collective behaviour, in which the individual's contribution to events is secondary to the social context. There is no doubt a lot of valuable insight gained in doing so, but the flip side of this is that the individual's relationship with society is the central factor of social dynamics, much as the atom is the core ingredient of molecular organisation. Accordingly, to understand developments in the domain of culture it's necessary to pay particular attention to the nature of the individual psyche.

A significant element of cultural organisation and behaviour, touched on in the last chapter, is 'identity'. This term relates to how a person perceives themself, and to how groups of people perceive themselves as collective entities. Since the individual attention is necessarily the active principle at the heart of both of these spheres, self-awareness is a central feature. It seems clear then that self-awareness contributes significantly to the dynamics of culture.

The question then arises, how is self-awareness represented in behaviour? My own observation leads me to focus here on 'self-esteem' as a particularly evident expression of self-awareness, widely represented in people's states of well-being and behaviour, individually and collectively. Also, while much of this discussion of culture and the place of the individual may appear rather abstract, self-esteem is an active point of reference that, with a little reflection, the reader may be able to relate to at a more personal level.

It's clear that self-awareness and identity are intimately connected. The term 'identity' relates to how a person perceives themself, and to how groups of people perceive themselves as collective entities. Since the individual attention is the active principle at their heart, both forms of identity originate in personal self-awareness and are completely interlinked. Accordingly, it seems clear that as a major feature of the psyche, self-awareness contributes significantly to the dynamics of culture. As I have observed earlier, self-awareness shouldn't be confused, or conflated, with consciousness, nevertheless, it seems fair to say that it is undoubtedly a major element of our human expression of consciousness and of our behaviour, personally and collectively.

A particularly prominent feature of identity is the sense of self-esteem. This is a term that can be characterised in different ways. Commonly, it's taken to mean something closely related to 'pride', and this view is reflected in the line from Burns quoted at the beginning of this piece. While appreciating the point he was making - and with no disrespect to the poet - I'll offer a somewhat expanded take on it here.

It seems to me that the concept of self-esteem refers to a sense of self-worth as a more primary, or original, quality of mind than 'pride' in the sense of ego. Pride in the latter sense relates very much to how we perceive other people to be perceiving us... For sure our sense of self-esteem is influenced by our relationship to those around us, but to my mind, 'ego', or reflective pride, is a kind of illusory by-product of insecure self-esteem.

In some respects, it can be easier to describe loss of self-esteem than self-esteem itself. An overriding quality favoured by evolution is equilibrium. In the context of human experience self-esteem reflects this condition of balance, and its absence generates pressure towards its restoration. In a society that functions reasonably in harmony with nature, self-esteem is a quality that comfortably reflects the relationship a person has with their community and is as natural and straightforward to sustain as walking on the ground. This condition can be seen as representing a fundamental sense of identity and social relationship from which our many and various cultures have developed and is represented to at least some degree in the myth of the Garden of Eden. But, the world we now inhabit and must find identity within is much more abstract and complex. Though the underlying patterns created by nature remain intrinsic, the challenges to maintaining self-esteem can be seen as increasingly comparable to the efforts of a tightrope walker to maintain balance. A few can do it well, most just get by, and some find it a great challenge.

Broadly, the need for self-esteem finds satisfaction in valid identity sustained through recognition by family, extended family and social peers, self-respect in behaviour, and fulfilment generally according to what is, or appears to be, socially desirable and available by circumstance, and at different stages of life. Bearing in mind its social context, in practice, self-esteem on these terms is largely associated with practical community issues. In the context of modern Western societies, it is generally gauged and sought with an emphasis on 'success', via career, material wealth, education, sport, and many practical forms of engagement or competition, or through direct peer recognition and choice of friends. It may be a natural result of constructive community involvement of many different forms, professional or otherwise. It may also be sought through the generally more subtle channels of music, art, drama, or other means of creative expression, as also through physical fitness, religion, self-inquiry, philosophy etc. Some of these avenues may seem rather trivial, but essentially, in addition to whatever useful purpose they serve, all are means of establishing one's 'fitness' and worthiness, to oneself, and in the eyes of the community in terms that community can relate to.

Where society has imbalances though, this appetite will often seek satisfaction in disturbed ways. In the absence of avenues for self-expression substitutes may be found or created in the immediate environment, sometimes creative and socially beneficial, sometimes destructive to the individual and the wider community. If we bear in mind that everything has roots at the level of the microcosm, the social microcosm is the immediate environment an individual inhabits and experiences directly on a day-to-day basis, family, friends, school, work, and economic circumstances. In this sense, the immediately prevailing environment and social circumstances, values, and conventions, demands and pressures, largely provide the context for self-esteem aspiration, the apparent dimensions of possibility and opportunity or lack of it. Consequently, they also create the perspective for perceived failure and loss of self-esteem. Clearly, circumstances such as economic and social conditions, including opportunity, play a considerable practical role in the background to the developments of peer culture and expectations. In large scale modern societies, the demands on self-esteem may be very much more complex and challenging than in earlier times but the same underlying principles and demands are present. Self-esteem may also be sought, not so much in terms of the social environment, but more internally, through self-exploration or testing by for example, pushing the limits of one's habitual boundaries of behaviour or skill. In other words, through challenging and proving oneself to oneself. In a sense this is always an aspect of self-esteem, however it is affirmed. Finally, but by no means least, at the personal level, self-esteem may be restored to at least some degree by simply re-establishing a sense of nature in one's lifestyle by various means, such as, for example, by spending time in natural settings where the complex demands of human society are less pressing, or by means of a simple meditation method, which can be very helpful towards restoring balance.

Self-esteem and identity are closely linked. If self-esteem, or the challenges it faces, are perceived and expressed in terms of the social environment, that 'environment' includes the social group or groups that we 'identify' with personally at a local scale, and more globally. Such 'groups' don't have strictly physical dimensions, more generally they are mental groupings of shared-mindedness, whether these take the form of political or religious ideology and belief, ethnic affiliation, or areas of technical or scientific specialisation, sporting affiliation, nationality and so on. All of these represent forms of what might be considered 'tribal' identification that we tend to adopt, through conditioning in the first place, reinforcing experience, indoctrination, circumstance, and habit in an ongoing sense. Many different forms of identification may be active simultaneously, overlapping, often with degrees of contradiction, and particularly in the age of the internet and social media, increasingly obscure. These regions of tribal/ group identification are spheres in which the tensions and aspirations of self-esteem become expressed.

Aspiration

In the view I am offering here, the factor of self-esteem is seen to be a considerable influence in social behaviour, reflecting the intimate relationship between the individual and society. As ever though, although personal self-awareness is at the core of this relationship, the environment provides the dominant sphere of influence and opportunity, and the directions explored by self-esteem reflect this.

If we consider the shaping influence that the environment has on individual behaviour and identity, a broadly common factor reflected across all cultures, can be found in the expression of values, discussed in the last chapter. Recognition of this is intrinsic within the very idea of 'culture', i.e., recognition that instinctive drives are refined and modified through collective preference, to the benefit (sometimes injury) of society at large. Again, it becomes clear that while 'culture' is generally thought about in static terms, at root, this concept reflects 'cultivation' as an ongoing process of development. The presence of evolutionary dynamic within the sphere of human behaviour is implicit here. In one form or another, the refinement of values is a result of 'aspiration', which as ever, has intimately related personal and collective dimensions.

In a sense, aspiration could be said to be a form of 'appetite' that appears to be unique to the human species and is closely associated to self-esteem. For sure, our experience of aspiration is very much influenced by and reflects, the abstract and complex nature of our mental environment, but it seems probable that the forms it takes are emergent from established appetites already present widely throughout life, rather than fundamentally different. In a sense, to 'aspire' is to seek to *be more, to grow*, to satisfy an instinctive urge or appetite of the psyche emergent from, and at root intimately related to, the appetites for food and reproduction, but *apparently* unique to the domain of human conscious experience. To begin with, aspiration generally revolves around basic requirements, for example, the need to satisfy or improve living conditions. This can be seen to apply to many spheres, from the creation of 'home', to social, political, and 'spiritual' ideals and desires. Aspiration can be seen as a natural extension of the basic appetites/instincts of biology - survival, sustenance, and reproduction - into the many dimensions of abstract attention that we increasingly inhabit, essentially, a subtle expression of the impulse to survive and propagate, essential for our personal well-being and social development.

Viewed in these terms, there is no fundamental difference between aspiration in the human context, and other forms of 'appetite' to be found throughout the organic domains, although there is clearly great difference in the forms generated. When we begin to recognise this underlying continuity, it could be taken to indicate that our many forms of aspiration are no more than subtly modified expressions of those basic instincts, remaining essentially basic. However, all forms of motivation, from those of elementary biology, through that of cognition and conscious experience generally, to that of human aspiration, are subject to the same principles of growing complexity, transformation, and refinement - in other words, hierarchical development. Accordingly, aspiration as an emergent form of 'appetite' of the human psyche is further expressed in more subtle spheres such as the desires for self-knowledge and understanding of reality, apparently far removed from the demands of metabolism, but still connected... These developments also are reflected in the broad spectrum of values, where transformation and refinement continue actively. Again, as with all matters of culture, aspiration has entwined personal and collective dimensions... Personal aspiration reflects the unique relationship the individual has with their environment, at the same time, aspirations resonant in the collective attention reflect the pressures accumulating there on account of their presence within many individuals.

Having made these observations about the intimate links between self-esteem, aspiration, and the social environment however, it seems to me that there is another important dimension to self-esteem. Self-esteem, or self-worth, is at root an innate quality. Another word for it is dignity, (which, incidentally, is by no means unique to human beings). A child is born having self-worth, not as a measurable or reflected quality, or aspiration, ego, or conditioning, but rather, as a sense of completeness that has never been put in doubt (this quality is one that we widely perceive in nature, often referred to as 'innocence'). This is a fundamental quality but fragile. The love the child normally receives from its' parents reinforces and protects it (as a rule). Only later is the security of self-worth challenged as the child engages more with the wider environment. Gradually, the individual develops a need to re-establish this sense of security by finding acceptance and fulfilment within society on society's terms, through channels that are available. At all stages this whole process is entirely natural and reflects the way in which we constantly find evolution proceeding in all spheres through the intimate involvement of individual entities with their environment. In this way however, the individual comes to seek self-esteem to a large degree in terms of the environment, tending to lose touch with their innate self-worth (again, this is identical to the idea of 'loss of innocence'). Aspiration too strongly identified with circumstance tends to create relatively superficial self-esteem, principally reflecting the views of others, creating vulnerability to manipulation and other forms of distortion but I won't explore this topic for now.

The search for self-esteem through behaviour that reflects superficial values and attitudes current in the environment alone is, by definition, bound to be essentially ephemeral, limited, and subject to the buffeting of circumstance. Instinctively, all cultures have traditions that serve the purpose of transmitting more significant value heritage from generation to generation. These take many different forms, from mythology to philosophy, drama to literature, religion to science. Such traditions help to sustain and protect society as it evolves by communicating insights born of experience that can assist the individual to identify socially and sustain self-esteem in positive ways. But the weight of doctrine and inertia in most, if not all, belief systems tends towards establishing views and rules to be accepted and obeyed with a minimum of reflection, rather than encouraging wider inquiry within freedom of thought. This latter, freedom of thought, appears to me to be an essential requirement for the sustenance and revitalisation of insight, self-identity, and social values of all kinds.*[1]

If, on the other hand, we consider the innate side of self-esteem it's possible to see that social conditions may assist in its expression or hinder it. This may appear contradictory at first glance; how can we conceive of an innate quality as something that can be encouraged or obstructed? In what way does that differ from imposing social norms and expectations? To my mind the difference reflects again, the relationship between 'nature' and 'nurture'. Nature provides us with all our fundamental features and qualities, as it does throughout all life. All living creatures are governed by such innate characteristics, long established by evolution, in both form and behaviour, but in the domain of human culture, we are able to organise our societies with considerable freedom to mould, or 'nurture', them according to our abstract ideas of how they should be, and of our relationship with the rest of nature. It's as if we are one with nature and at the same time stand outside of it, viewing it as observers, expressing and enacting interpretations and opinions. Accordingly, the social environments we create can be more, or less, representative of our evolutionary heritage, with profound consequences for our mental well-being. As far as both the individual and the society are concerned the living source of the expression of this innate aspect, clearly, is within the individual, but the social forms we conceive of and create *may* express and contribute to its sustenance or obstruct it. An image that represents this relationship is that of plants growing in good soil, with the presence also of air, water and sunlight. Where all of a plant's requirements for sustenance are available, its innate qualities are expressed spontaneously and effortlessly, but where any of these environmental conditions aren't fully met the plant can't develop to its full potential and is likely to grow weak, distorted or stunted, even though its innate qualities and potential continue to exist, intact but poorly expressed. It seems to me that in Western societies we have gone through a long period - many centuries now - in which understanding of this balance has been greatly obscured, but never totally, and the discomfort that results generates a desire to correct the deficits in the environment. I would see this impulse as having its roots in the same evolutionary balancing principles to be seen throughout nature that I constantly refer to. In that sense, the whole situation - of having the freedom to lose sight of our deeper roots, to become aware of that, and to consciously act to restore connection with them - are inclusive aspects of our evolutionary experience and growth, to be recognised and understood as an integral feature within our attitudes towards social organisation.

Recognition of the importance of individual discrimination is no doubt reflected in the current growth in aspirations towards personal freedoms and liberal mores, but attention still tends to be directed towards the external features of social rules, rather than internal ones. In other words, on attempting to reduce the restrictions of social rules, but with less attention paid to the much more intimate restrictions of internal conditioning and identifications which can only be dealt with internally, i.e., through self-inquiry. In modern times, perhaps mainly because of the pressures created by large, complex populations, there is an understandable tendency to think of such matters in terms of social engineering, that is, in terms of economy and all that entails, political ideals, policy etc. which puts the focus of attention on material circumstance and the forms of social organisation. From that perspective the condition and well-being of the individual is generally viewed as a result rather than a cause of social conditions. Clearly there is a lot of truth in that but it's not the whole story. If, on the other hand, one takes a perhaps more traditional view, that the well-being of the individual is largely in their own hands (and God or providence, however conceived…), there is no reason for social progress, no injustice or inequality. In a sense this has been implicit in the teachings of some of the major religions for many centuries; all is ultimately in God's will, so whatever happens accept it. Viz, 'give unto God what is God's, unto Caesar what is Caesar's'. Clearly a very pragmatic view for feudal life, or under dictatorship, or empire; less suitable though for the ethos of democracy. There seems to be very little an individual can do to bring about significant change in their environment, accordingly, religion is viewed primarily in terms of personal 'spiritual' well-being; politics, in terms of 'material', social matters, as if the two spheres were entirely separate.

The emergence of democracy has helped to shift this perspective considerably however, by raising the sense of engagement of the individual with community circumstance, and enhancing, if gradually, the sense of personal empowerment, and the responsibility that entails. In that sense, democracy is not only a means of satisfying the aspiration to personal involvement in community decision-making but contributes to the direction collective identity develops. The point here is that democracy isn't a destination, rather, it's a doorway to further developments. I'll digress here briefly to discuss this topic a little more.

Democracy

Democracy can be seen as a blunt tool to regulate the competition of views and opinions by means of collectively acceptable compromise, rather than an instrument of true co-operation; certainly, how it is currently represented in much of the world fits that characterisation. Still, it establishes the principle of co-operation in human affairs as having priority over competition...

The *ideal* of democracy reflects a general recognition that the well-being of an individual person in a community depends on that of all the others. This recognition is a more or less innate feature of all family/tribal cultures, but in larger, more complex societies, for it to be satisfied, the individual has to able to participate directly in collective decision making, reaching beyond the tribal mentality. Democracy is a method to achieve this that has evolved/emerged through the collective attention, internationally and over centuries. Here, in principle, each person contributes to a balanced society, while a society that can work together to progress in co-operation provides a supportive environment for the individual to grow and develop their potential. But this represents the aspiration at the heart of the ideal, not a description of its current status. While democracy allows for greater self-expression, the central instrument is really individual self-interest and this doesn't necessarily benefit community well-being in the short term. Nevertheless, the aspiration carries great significance in my view, because it opens the way to the development of a greater sense of community participation and responsibility on the part of the individual, and over time, an improving balance of self-interest with collective well-being - not through planning, but through the natural developments of competition and cooperation within the social sphere. This aspiration is more important than the concept 'democracy', which simply reflects it - though that is important too; it represents a tool we have created towards realising the aspiration.

Both sides of this relationship are important and require attention. It could be reasonably said then, that approaches of social organisation should actively support self-development as a basic requirement of balanced society. In modern secular societies this is increasingly the case but still largely gauged in material terms, particularly economics, which is intensely competition oriented. The result may be improved focus on basic practical needs, but doesn't provide much support for more subtle dimensions of personal fulfilment, community coherence and self-esteem. Such matters used to be addressed by religion, and for many this remains the case. For more secular societies, though, the challenge lies in finding a meaningful basis for engagement with profound realities of the human psyche, that can bridge between present day insights into the workings of nature, and older world views, across cultures, without discarding timeless wisdom such views may contain, particularly since modern day understandings are increasingly based on essentially technical principles. These may be useful in practical spheres but contribute nothing, to moral or ethical considerations which remain vital to community well-being, other than through shifting social pressures they generate.

There have of course been many attempts made towards creating ideal societies - most of which have been based on defining and controlling the behaviour and beliefs of the individual - at different times and in different parts of the world, with varying degrees of success. I'm not going to embark here on a close look at these, I'll just observe that attempts to impose a fixed type of social ideology on a population, by means political or religious, don't generally appear to advance societies much, and have rarely been to the advantage of 'ordinary' people, more often, and at all scales, to a self-selected minority. In contrast, social forms that engage and encourage individual participation creatively and in decision making to some degree, tend to thrive, emotionally at least, if not necessarily in our contemporary terms of economic productivity. Beyond essential basic needs, social harmony appears to be more important for personal wellbeing and self-esteem than material wealth per se. This applies to both larger modern and smaller, more traditional societies, though in somewhat different ways; the more traditional ones tending to be, you might say, more 'organically' based around extended family relationships and hierarchy, while among the larger scale modern societies democracy is emerging as an expression of the aspiration towards fuller engagement of the individual at the heart of collective decision making, as this is where needs originate. Democracy is a significant development within the domain of culture, but even so, at this stage the individual is largely still viewed as a cog in the processes of political organisation, rather than the core ingredient. Their role is largely limited to making what amounts to a token contribution periodically, in the form of a 'multiple choice' vote, towards the decisions of others better positioned to bear influence, rather than as a genuine informed participant in decision-making. Much work to be done here...

The emergence of democracy as an ideal is no doubt an important landmark in the domain of social evolution, but it's not just something that you have or don't have. Like everything else, 'democracy' must evolve, and the course of that evolution, as a consciously emergent social principle, lies in the care and attention we put into it, not an assumption that it is a fait accompli, which will automatically transform older outdated, approaches to organisation and control. It requires *cultivation* to grow from a young and fragile ideal to a mature condition that really represents the ideals it aspires to. In these terms, democracy is a significant ideal but still at an early stage of development. Still, it emerged 'through our hands' and it is with 'these hands' that it must be cultivated.

Returning to the theme of self-esteem... In all evolutionary contexts, competition is an important element in the refinement of developments and appears as an unavoidable, frequently enjoyable, and essential aspect of everyday life. But where the element of competition and its effects become dominant factors in the fabric of society this can have a great impact in shaping the orientation of self-worth and dignity in terms of material achievement. One result of this is a tendency to link 'self-worth' strongly to social status, based on one's perceived value to society in 'material' terms, encouraging a kind of ego scale of self-worth that reflects this evaluation. This may be to some degree unavoidable, but that doesn't mean that this is the only or best way to focus and engage the aspirations of people, particularly as it feeds very early into the ethos of education. Many other impulses contribute significantly and more fundamentally to the quality of life and ability to attain personal satisfaction and fulfilment, such as personal relationships, family, friendships, community participation in all its forms - i.e., the more emotionally cooperative dimensions of our existence. It's just that many of these tend to be under-valued, under-resourced and simply taken for granted when the attention of society is largely directed in a manner involving strong elements of competition, particularly the modern system of economy and the resulting working relationships, job security, social status etc.

It seems to me that emotional security is an essential support of innate self-esteem, and again, a basic feature of the psyche. It is first and foremost a measure of emotional balance and well-being. Material success doesn't bring it and failure doesn't necessarily undermine or damage it unless it is already fragile. Also, it seems a reasonable observation that those who are emotionally balanced, secure, and confident in themselves are more likely than those who aren't, to concentrate on and succeed in activities they engage in, whether personally fulfilling or just required by circumstance. Accordingly, these qualities contribute to community in supportive ways of all kinds, emotional and economic, technical, and aesthetic. This of course isn't a new observation, and educational approaches have been developed to some degree that acknowledge it, and encourage a holistic attitude towards children's learning. But this has scarcely filtered into mainstream attitudes, which are generally dominated by the focus on more immediate and demanding economic and related concerns. This isn't to underestimate the importance of such matters, but perhaps to question the completeness of an approach that tends to encourage albeit inadvertently, a sense of insecurity as a basis for creating security...

Amongst the practicalities of social organisation, education occupies a special place, particularly in Western cultures, where religious ideology has been giving way to broader, more secular views on personal development for some time. In many ways this is undoubtedly beneficial, for reasons I have been discussing regarding the growing pressure towards personal discrimination, broadening the mind at the same time. But secular education, except perhaps in a very few societies, currently tends to be weighted towards practicalities of existence with an implicit emphasis still, on competition, with a minimum of attention toward the sphere of values and self-development with the assumption that this is largely beyond its remit, traditionally more regarded as that of religion.

The major world religions have generally used their authority (asserting divine inspiration) to bring to attention and back up the esteem with which many values such as honesty, fairness, compassion, respect for life and nature are held in societies, emphasising their natural (i.e., divine) origin, but meanwhile, asserting that they can only be established through indoctrination. Finding a way of maintaining that sense of esteem in a secular context requires in the first place, going back to basics to recognise that these qualities don't have to be acquired or instilled but already exist as innate qualities of the psyche and finding approaches to their sustenance and cultivation that don't rely on this form of conditioning.

Moving towards an outlook that pays more attention to the requirements of personal emotional security and development would seem to be not only an attractive idea, but an essential one for the health of society in all ways - including economic ones. If the integrity of the society depends on the integrity of the individual, encouraging individuals to explore their own personal abilities, interests, and skills, and develop in self-confidence and outlook, must be better than obliging them to compete strictly within defined regimes - resulting in 'failure' or mediocrity for many, perhaps a majority. No doubt a better balance could be aimed for, where in the longer term the individual may be more inclined to make responsible and appropriate practical life decisions from a sounder basis in self-confidence (I don't pretend to be saying anything original here, I just want to emphasise the importance of the emotional well-being of the individual in the balance, over considerations conceived of in terms of economics).

The link between self-esteem and social values is profound and dynamic, an invisible thread of social fabric as-it-were, and is at the centre of social integrity and collective well-being. The positive side of this relationship is found in relatively harmonious community organisation, where collective responsibility and engagement naturally enhance co-operation. However, these are just observations from personal experience. It is possible to observe how nature works, and be enlightened thereby to some degree, but I believe that attempting to translate perceptions of this kind into ideology is intrinsically flawed. If anything, we interfere with nature too much, the challenge is to learn how to provide support but interfere less. The negative side can be seen in the growth of social problems such as social deprivation (commonly referred to as poverty, but this term really relates much more broadly to restricted opportunity and aspiration than just to immediate financial conditions), alcohol and drug abuse/addiction, and mental stress disturbances as well as anxiety and depression. These should be recognised primarily as symptoms rather than causes, of loss of social integrity. One way of looking at this is to compare the situation with that of environmental health, where ill health and disease can develop and proliferate because of poor basic conditions and ignorance of hygiene.

Although specific problems must be dealt with directly, and 'contagion' is an element in the context of both disease and social behaviour, if imbalance in the underlying conditions remains, the problems can only continue to present. Being hierarchically framed, the 'causes', and potential cures, of endemic social imbalance originate more at the higher levels of social organisation than at the lower levels where the symptoms appear most obviously. Where the basic conditions of social sustenance and 'ecology' are poorly supported, poor community health, physical and mental, is not surprising...

These observations just begin to touch on the relationship between an individual person and their society. The key point I want to make here is that self-esteem is a central factor in that relationship, intimately woven into the forms that identity takes, personally and collectively.

Fundamental to self-esteem is of course, self-awareness. This may appear to be a very abstract phenomenon, difficult to consider in practical terms, but it is an evolved, emergent, feature of our overall make-up and can be understood to at least some degree in terms of our nervous system. I'll be discussing this topic further over the next two chapters.

Summary

The basis of culture is society - which is, people relating with and to one another. This relationship is the basis of what we call identity, and at the heart of identity, both personal and collective, lies self-awareness. An active element in this relationship is the sense of self-esteem, the value in which a person, and a society, holds themselves.

This is intimately tied in with the sense of self-confidence and that of security. A child is born with an innate endowment of these qualities as the heritage of evolution, but they are necessarily linked also to the environment that is inhabited. In its own way, this reflects an evolutionary imperative within the domain of culture. In a sense, self-esteem is a composite term, reflecting a range of factors concerning the well-being of the psyche. In practice, it can be - and frequently is - developed in superficial and imbalanced ways that direct emphasis more towards either the emotions or the ego, in ways that reflect and reinforce tendencies predominant in the environment. For which reason, insecurity may find expression in excesses in either of these directions, ranging from a sense of inadequacy, to excessive empathy, to the desire for material superiority, or to control others.

Ideally though, self-esteem reflects a balanced relationship between these different aspects of the psyche. The demands of evolution are such that we are required and obliged to participate in this dynamic, but there is also good reason to suppose that, for the most part, the centre of gravity of the balance of the psyche remains within the jurisdiction of our evolutionary heritage, not that of the relatively superficial dimensions of circumstance. Most societies appear to tacitly recognise this and reflect this recognition through teachings of various forms, such as mythology, religion, and philosophy.

Chapter Note

*1 in the last chapter I observed that values represent rules of behaviour that assist freedom of movement at the collective level. Language, and how it is used, plays a major role in this clearly, ideally facilitating the free movement/exchange of ideas and information, but this isn't the case when rules become dogma. In my view rules for their own sake interfere with the perception of truth and growth of insight instead of assisting.

Chapter 10

Self-awareness and the Nervous System

Outline

I've discussed human culture as the domain currently at the active peak of the evolutionary hierarchy, I've considered the position of the individual, and I've touched on self-awareness as a form of experience significant in the dynamics of this domain. In the last chapter I discussed the 'form' of self-awareness and identity as reflecting to a considerable extent, the impressions of others, and observed that this has effects at the collective level also. Again, there is nothing original in this observation, but the context here is that of evolution, which offers fresh insights as a way of thinking about it. Self-awareness now plays a major part in events and may play a significant role in ongoing developments within the field of consciousness.

This chapter considers the nature of self-awareness in a little more practical detail. The form it takes reflects our psychological and physiological makeup and their common evolutionary origins. It is also greatly influenced by the environment, nevertheless the primary qualities it reflects are inner features of the psyche, and these can be understood to at least some degree in terms of the nervous system. However one conceives of consciousness and self-awareness, clearly the nervous system is the instrument of their expression, accordingly, its nature is the focus of this chapter.

The way in which the human nervous system is structured reflects patterns in the organisation of life that can be recognised as occurring to some degree at all levels. Regardless how simple or complex the life form, these basic principles of existence remain essentially the same and must be satisfied for any organism to survive and prosper. Accordingly, the nervous system, in all the forms found in different organisms, finds its roots in the motivations and functions of cellular existence, the most basic of which are the requirements of cell metabolism.

'Metabolism' refers to the central processes of cellular function, the intake and assimilation of nutrients, and consequently, the elimination of the resulting waste products. While for the most part, single cell organisms depend on *relatively* simple and direct methods of satisfying these requirements, multicellular organisms have evolved a variety of means through which they are handled on behalf of the entire organism by devolving aspects of the task to specialised sub-structures. (Many single cell organisms also do this to a degree, but in a less complex manner.) The nervous system represents an advanced level of specialisation of this kind. Not surprisingly then, in the simplest multicellular organisms that we can currently observe that have one, such as nematode worms for example, the nervous system is largely focused on the purposes of acquisition of food, digestive and reproductive functions and motion control towards those ends.*[1] As organisms become more elaborate, layers of complexity are added to the dimensions of function, yet these core features remain fundamental and central. If we consider the human nervous system against this background, certain aspects of its specialised structure can be recognised, and some evolutionary context offered.

Overall, the human nervous system comprises several integrated spheres of function, which together enable and regulate

our activities as an organism. These are ongoing results of continuous evolutionary process over the period of life on earth and
therefore represent and reflect the nature of that process and most
if not all its significant stages in its form. The view I offer here
may seem a little unconventional in that it draws partly on
Western descriptions of the physiology of the human nervous
system, partly on descriptions originating in other cultures, which
focus more on the sphere of conscious experience.

In the terms I have been outlining, of hierarchical organisation, the development of the nervous system has been taking
place under the jurisdiction of cognition since at least the
emergence of the first form of nervous system that we can reasonably identify, some 550 – 580 million years ago at the beginning
of the Cambrian epoch, when complex multicellular life forms
generally were developing in great variety. (This, in turn, was
made possible by the much earlier emergence of the eukaryotic
cell structure, touched on in Chapter 4.) Among other significant
consequences for the forms that life has taken and that surround
us today, the emergence of the nervous system was a major development from that stage of events, representing the coordinated
control of the diverse cellular processes required by a complex
organism. This development appears to have taken place quite
quickly in terms of evolutionary time scales after the appearance
of the earliest forms of multi-cellular organisms.

In governing the supply and maintenance requirements of
metabolism, the nervous system exercises the broad functions of
experience and *response*. For this reason, although the end results
are clearly physical and dedicated towards the purposes of sustenance and reproduction, its most significant sphere of function is

cognition, from the simplest expressions in nature that we know of, to the most complex, and with some mutual common ancestor somewhere in the ancient pre-Cambrian era. Accordingly, the nervous system needs to be considered not only in strictly physical terms but also in terms of the dynamics and mental structures of cognition.

The human nervous system is incredibly complex, by far the most complex structure in our known universe, still, it has many features that are recognisable as being widely represented in nature. These have been explored to some degree both via the physical methods of Western science, and in Eastern cultures through very different approaches involving internal, mental, inquiry and exploration. Despite the apparently great difference in these approaches there are some distinct parallels in the descriptions that have resulted. One particular aspect of the central nervous system stands out as relevant to consider here. This is referred to as the *autonomic nervous system* and is the main subject of the following discussion. This is the region of the system that supervises, coordinates, and maintains the normal balanced functioning of our organism, including all its constituent parts and organs, from the level of metabolism upwards. At the same time, it mediates our engagement with the environment, adjusting the state of all these functions according to the demands of circumstance. It comprises two main aspects referred to as the *parasympathetic* and *sympathetic* systems, which work together to regulate both our physical and mental conditions. The parasympathetic functions as what may be regarded as the reservoir of evolutionary integrity, constantly seeking to return the functions of our organism to a condition of metabolic balance as established by our genetic history. Meanwhile, as the name suggests, the sympathetic system functions to modulate the state of the organism according to the challenges and demands of everyday circumstance. (Commonly, the function of the sympathetic system is characterised in terms of the 'fight or flight' response, but it does much more subtle work than that suggests, as it constantly makes fine adjustments to our physical and mental conditions. The 'fight or flight' condition represents only an extreme state of this response.) The parasympathetic restores 'normal' metabolic balance, the sympathetic aspect modifies, or modulates, it.

The sympathetic system is further divided into left and right regions, or aspects, at the level of the brain, and as a network of distribution within the organism. These two spheres of cognition operate together in a complementary way, effectively as an instrument of the psyche for the modification of the state of the physiology. At the level of the brain, which has a governing role in coordination, the cognitive aspects of these functions, loosely summarised, are; predominantly emotional experience and memory for the right sphere, action, planning and rationality for the left. The locations of their physical representation in the body are reversed, as these aspects of the nervous system cross over outwith the brain.

The autonomic system, as a whole then, functions to maintain and regulate the integrity of the organism, while mediating its engagement with the environment, adjusting its state accordingly, from the level of metabolism to that of overall behaviour. The baseline of this engagement is physical of course, but it necessarily involves our mental sphere also. This includes all our emotional and rational aspects - you might say, our psychological dimensions. In practice these are intimately entwined with our social environment, which means that the mental environment effectively plays just as real a part in our condition as the physical. An obvious consequence of this is that the pressures and demands of complex social conditions have intimate and sustained effects on our mental *and* physical states. Altogether, the autonomic system offers a considerable source of insight into how we function in the present moment, and about our evolutionary history, as it directly reflects the evolution of motivation and action throughout the entire domain of cognition.

Darwin and Wallace illuminated much of the mechanics of evolution, but evolution has been central within the Indian philosophic view of existence for millennia, though perceived more as a continuum of 'being' in which all living things participate, rather than physical attributes. Western descriptions of the nervous system derive largely from surgical methods of inquiry and are couched mainly in physical terms such as neurons, nerve fibres and plexuses, centring on the brain and its functions. Indian descriptions of the nervous, or subtle, system as it is known, derive from internal mental explorations in the form of meditation methods rather than physical ones. Such descriptions tend to portray the relationship of mental and physical states in a more integrated way, but with priority towards the sphere and dimensions of experience. Aspects of this relationship are highlighted and associated with certain focal points of the autonomic system distributed throughout the body, with various spheres of function identified and described.

These spheres of function concern both physical and mental qualities, but primarily the latter, as, the dimensions of experience are regarded as principal. Focal areas of the nervous system correspond broadly to the aforementioned plexuses and are traditionally referred to in India as subtle centres, or 'chakras' (which many readers may have heard of. This knowledge derives from a long and respected tradition of inquiry in India but has been tarnished somewhat by misunderstanding, misuse, and exploitation, particularly, but not only, in the West, in recent decades). These centres are described as having characteristics relating to their contribution to both the dimensions of mental experience in which they participate and associated aspects of physiology. They are generally represented as being located in seven principal locations in the body, with many sub plexuses, and are mainly associated with the spinal column and the brain: 5 in the region of the spinal column, 2 in the brain, and as having wider correspondence with universal 'qualities' of nature. They represent the major points of contact between our established evolutionary roots - via the parasympathetic aspect - and the sphere of our immediate engagement with the environment, emotionally and physically - via the sympathetic aspects.

The qualities these centres express are considered to reflect universal evolutionary principles that govern all the major stages of development that have occurred since the 'beginning of the universe' - however we may conceive of that - and to underlie all levels of natural form.

This kind of description presents a view of the human organism in which the evolution of the psyche is the focus of attention rather than physiology, and in which, as I have discussed in earlier chapters, the laws of nature we can identify in purely physical terms represent just the lower levels of evolutionary development. The perspective of evolution in which this view rests differs considerably from the Darwinian version we are generally more familiar with, where attention is primarily on technical considerations such as natural selection and ecology, nevertheless, there are many points of correspondence, especially when hierarchy is considered. The approach of Western inquiry into evolution has only been equipped to proceed in the manner it has since the invention of the microscope and telescope effectively, and the loosening of the hold of theological dictat over the ethos of inquiry, all of which has taken place relatively recently - over the last three or four centuries. Eastern inquiry into our relationship with nature via the direct exploration of conscious experience, has been ongoing continuously over several thousand years - with much more freedom of thought, and the perceptions expressed therein reflect a broad range of insight. This inquiry has contributed profound understandings and descriptions of the nature of human existence that shouldn't be underestimated in the face of modern, more technical approaches. These include descriptions of our intimate relationship with the rest of nature, of the holistic nature of our innate psychological condition, and of the way in which fundamental principles of evolution are expressed in all aspects of our form. The central ideas I present here derive from India, particularly from the insights of the Advaita Vedanta philosophic tradition and yoga, and especially, those of Sri Mataji Nirmala Devi Srivastava,*2 to whom this book is dedicated. The next few paragraphs present a brief outline of the view of subtle centres/chakras.

While, from the Western biological perspective the brain is considered to be the seat of consciousness and controller of the nervous system, which acts as its extension throughout the body, the 'chakra' system views the entire nervous system as one integrated entity, referred to as 'the subtle system', comprising several principal focal regions of coordination located in the spinal column, culminating in the brain. Each of these regions is understood to actively represent a stage in the deep history of the evolution of body and mind, the point of contact, control and coordination between them. But at the same time, having evolved together as one, 'body' and 'mind' are understood to be inextricably interlinked rather than entirely distinct qualities.

The chakras therefore, aren't described in biological terms, but rather, in a way perhaps more akin to that represented by the idea of archetypes which we find in Western psychology (courtesy of Carl Jung). Descriptions deriving from internal, mental, exploration tend to be personalised whichever culture creates them. This is a point at which descriptions physical and metaphysical tend to diverge. From the perspective of Eastern philosophy these 'personalisations' represent principles of nature reflected broadly across living nature, and in terms that reflect widely recognisable relationships, particularly those of parents and their offspring. At the level of human behaviour these images are used to represent the expression of these principles in the make-up of the human psyche.*³

Natural principles that can be observed at both macroscopic and microscopic scales of events are considered to be represented here, i.e., at the scale of universal principles and at the scale of human experience (reflected also in the expression, 'man is made in the image of God'). From a Western perspective, in the absence of insight into the symbology, such personifications have been generally viewed as diverse 'god' figures, and characterised as 'polytheism'. At least as represented through the more profound

Eastern religions and philosophies, their function can perhaps be better understood as symbolic representations of diverse but integrated primordial aspects, or qualities, of 'universal being'*4 rather than as individual 'gods'. Their origins lie in cultures and times much less complex than ours now are, much closer to direct experience, less overlaid with conceptual abstraction. From such a background, meditative exploration over several thousand years has been primarily concerned with the inner nature and condition of the 'observer' rather than the attending circumstances. Even in the cultures within which these perceptions and ideas developed however, there can be considerable differences in how they are interpreted and expressed. The outline I offer here is based on a particular interpretation that I have some familiarity and experience of and have personally found to be profound and holistic.

While Western science and its supporting philosophical approach focuses on the 'material', or 'objective' aspects of reality then, the Eastern focus is primarily on the dimensions of 'experience' and the nature of the experiencer. From this perspective, the human nervous system is regarded as an evolved and evolving expression of, and window upon, the entirety of evolutionary history, expressed through the medium of consciousness, with each of these seven centres representing and expressing, a particular facet of evolutionary dominion. (The uppermost of these centres is regarded as ultimately containing and integrating the others, in a way that can be compared to a head office that presides over and governs branch offices). These facets are not considered to be purely mechanical, or physical (or deterministic, incidentally), but rather, to express dimensions of primordial, universal consciousness, or 'being' which underlie and bear influence upon the patterns of evolutionary self-organisation in all domains. As such, this influence is entirely 'internal', shaping form via the instrumentation of selection in the environment according to 'rules' that emerge in particular domains. In that sense, the influence of these centres could be compared to that of 'fields',*5 expressed in the various evolutionary domains according to the rules and boundaries of those domains as created and established by prior evolutionary developments. For example, at the level of purely physical processes, these rules are known to us as the laws of physics and chemistry; at the organic level we conceive of them in terms of biology, genetics, and ecology; in the domain of human culture, we find them represented as values of social behaviour, which I have discussed. (Eastern philosophies such as those of Buddhism and Hinduism refer to these rules, throughout all levels, as 'dharma', which can be approximately interpreted as 'natural law'.)

Despite the very different forms of inquiry into nature engaged in by the cultures of the East and West, many interesting similarities have resulted, including descriptions of the structure of the nervous system, and recognition of the evolutionary nature of existence, but at the same time there is a very particular difference. The Western objective empirical approach, as it stands currently, commits us to understanding and interpreting evolution in strictly physical terms, excluding as far as possible the role of the observer in its deductions and descriptions, consciousness being regarded as entirely phenomenal and secondary, while the Eastern approach focusses much more on the nature and role of the experiencer, or observer, as central to our understanding and description of nature, reality, and everything those terms include.

When viewed with an understanding of their links to the

nervous system, it is easier to recognise that the more metaphysical and mythical descriptions and symbolism of Eastern religions and philosophy articulate profound perceptions about the nature of conscious experience itself, both in the individual and in nature more broadly. These descriptions have many links and parallels with Western ideas and mythologies, partly no doubt because they express common perceptions and insight, but often also because, for reasons of wide cultural transmission and exchange of ideas over thousands of years, many such perceptions have percolated from Eastern culture to the West. There is a tendency of longstanding in the West to believe that modern 'civilisation' and its important insights and discoveries began in Europe, but many of our most fundamental ideas, scientific, religious, and philosophic had their origins in distant cultures and times particularly, though not exclusively, the Middle East, India, and China.

The Sympathetic and Parasympathetic Aspects

The terms I will be using here serve the purpose of illustrating certain views of how the human psyche functions but have no, or very little, empirical scientific basis as such. Of course, this caveat applies to virtually all inquiry into the psyche, the study of psychology notwithstanding. From the perspective of science and its methods, we know much more about the physical universe than we do about the workings of the human mind, or any form of mind for that matter - you might say - as it has - more about outer space than inner. As I've indicated, the following understandings derive mainly from Eastern explorations and descriptions, but as far as possible I try to use Western terminology that suits.

The accumulation of 'past experience' in one form or

another, is a fundamental feature of evolution, and underlies the directions that all development takes. We see this expressed most clearly in the forms of genetic organisation, which represent this principle physically for organic and underlying molecular processes. It's also clear that past conscious experience is a major contributor to the mental forms developed by the individual and in the social environment. Within the individual psyche this aspect of the human attention, past experience, is represented by the subconscious, via the instrumentality of the left side sympathetic system. The most obvious expression of this faculty is of course memory, but as I've touched on earlier, accessible memory represents only the surface of the labyrinth of conditioning that is laid down over time and experience throughout a lifetime. The sphere of the psyche governing the left sympathetic resides (mainly) in the right side of the brain. Its dominion is primarily that of the emotions and is referred to as the 'superego'.

Meanwhile, the functioning of the right sympathetic concerns action. If one considers the 'character' of the left sympathetic as being strongly associated with past experience, and emotional conditioning, that of the right side is anticipation of the future, with an innate character of action (and attendant planning). This disposition of action of the right sympathetic is also subject to the development of habitual patterns of behaviour, but the sphere of the psyche this tendency develops is not so much that of the superego but of 'ego', which to a great extent associates self-identity with the 'fruits of action'. This description is not intended to suggest that these regions function independently, or that their partnership is simple.

The Sympathetic System

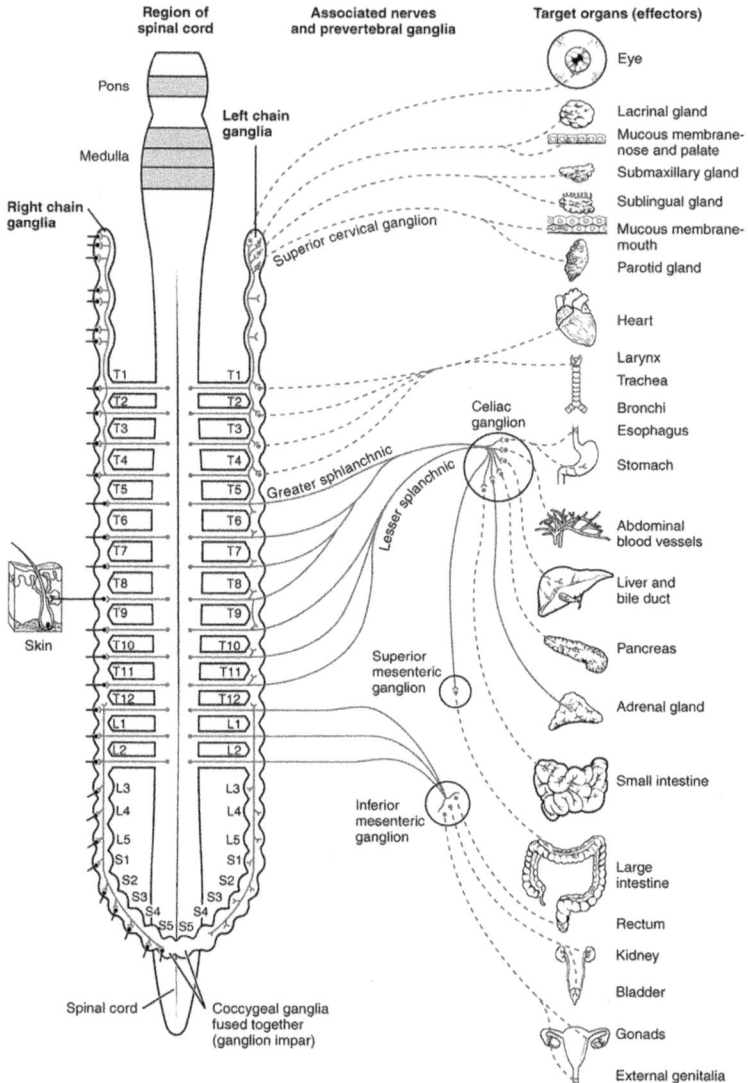

Source:
OpenStax College - Anatomy & Physiology, Connexions Website.
http://cnx.org/content/col11496/1.6/, Jun 19, 2013

The Parasympathetic System

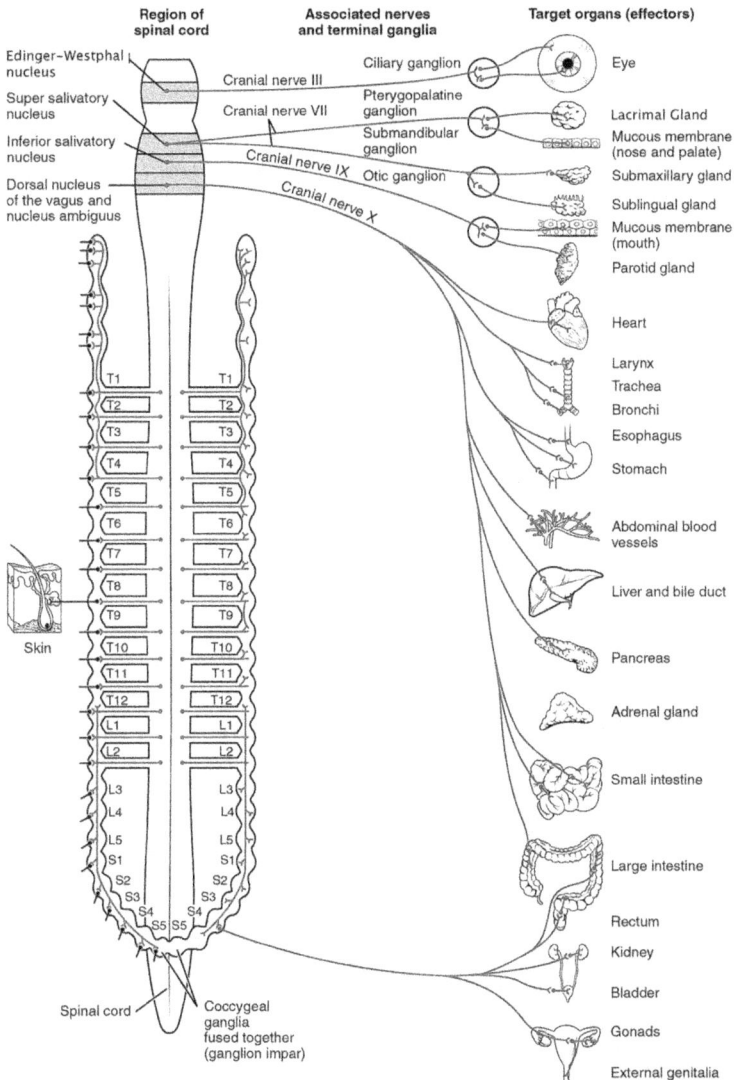

Region of spinal cord

Edinger–Westphal nucleus

Super salivatory nucleus

Inferior salivatory nucleus

Dorsal nucleus of the vagus and nucleus ambiguus

Skin

T1, T2, T3, T4, T5, T6, T7, T8, T9, T10, T11, T12, L1, L2, L3, L4, L5, S1, S2, S3, S4, S5

Spinal cord

Coccygeal ganglia fused together (ganglion impar)

Associated nerves and terminal ganglia

Cranial nerve III — Ciliary ganglion

Cranial nerve VII — Pterygopalatine ganglion

Submandibular ganglion

Cranial nerve IX — Otic ganglion

Cranial nerve X

Target organs (effectors)

Eye

Lacrimal Gland

Mucous membrane (nose and palate)

Submaxillary gland

Sublingual gland

Mucous membrane (mouth)

Parotid gland

Heart

Larynx

Trachea

Bronchi

Esophagus

Stomach

Abdominal blood vessels

Liver and bile duct

Pancreas

Adrenal gland

Small intestine

Large intestine

Rectum

Kidney

Bladder

Gonads

External genitalia

Source:
OpenStax College - Anatomy & Physiology, Connexions Website.
http://cnx.org/content/col11496/1.6/, Jun 19, 2013

The Subtle System

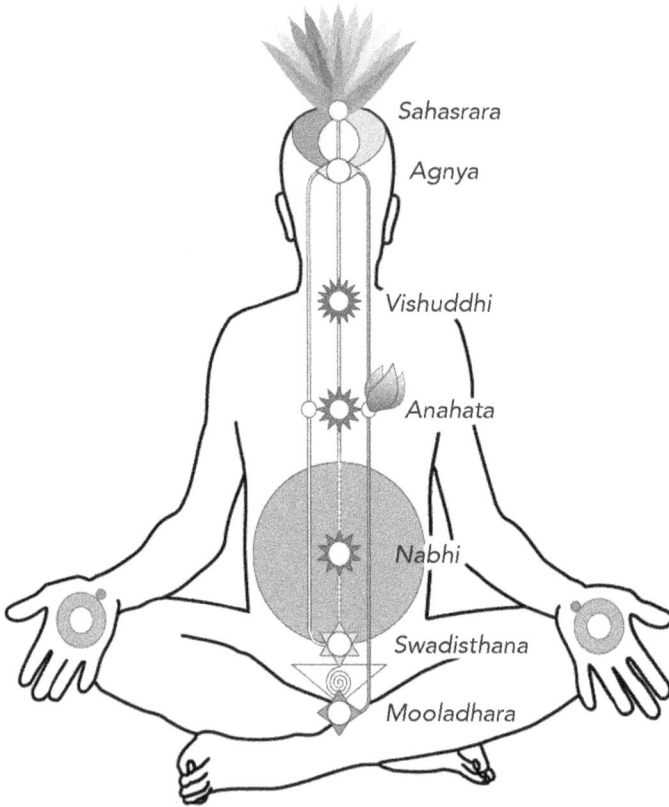

This image is based on the traditional Indian understanding of the human subtle nervous system. Principally, this comprises 7 'chakras', or subtle centres, and 3 'nadis', or channels. These correspond to the principle branches of the autonomic nervous system, with the parasympathetic in the centre and the left and right sympathetic branches facing the viewer. Here, the central channel is known as the Sushumna nadi, the left sympathetic as the Ida nadi, the right sympathetic as the Pingala nadi.

(Image courtesy of Patrick Anslow)

Nevertheless, the patterns generated through this relationship of 'motivation' to 'action' is represented in all life as we know it and finds expression in many ways that can be recognised. At the level of the human psyche these two spheres, ego, and superego, normally dominate the attention, providing the main features of self-awareness and identity. While they originate within the individual psyche, these features are reflected in all collective behaviours also. Although they appear to be expressed to a much more complex degree in our conscious field of attention than in other organisms and play a prominent role in the human psyche, nevertheless this combination of underlying qualities, can be seen to be a basic feature of the organisation of the nervous system of all organisms that we have some knowledge of that possess one. The relationship can be traced back to the earliest stages of complex animate life, and beyond that ultimately, to the realm of cell metabolism. Underlying and unifying subconscious memory, emotion, and conscious action are the same imperatives that drive all living organisms, simple or complex. According to the kind of organism we observe, we may characterise this motivation in different ways, but it can be generally translated as the impulse to survive, sustain, reproduce, which is expressed in 'higher' life-forms in the form of appetites. In the context of the human psyche this impulse is a principal factor that we might broadly refer to as 'desire', which takes many different forms of expression, from the most basic physical demands to the most subtle aesthetic sensitivities. Viewed in an evolutionary sense, human desire is on a continuum with the motivations present throughout life at all levels. We tend to characterise 'desire' as somehow unique to our species, but this distinction derives from older (largely Western) views, in which we were unaware of our continuous linkage with the rest of nature. There is still a tendency to think in this way, but the more we inquire, the more do such boundaries of distinction tend to disappear.

From the level of cell metabolism to the full range of conscious experience, one might say that the term 'desire' represents the driving impulse in all the domains of life. Survival, sustenance, and reproduction represent its foundations at the level of biology and continue to broadly underlie most human behaviour. However, it may be a mistake to conceive of this in strictly biological terms. Remember, *hierarchy*; within the sphere of human conscious experience, these terms take on a more subtle meaning in which the sustenance and propagation of ideas and insights contribute to the ongoing developments and fertility of culture. Accordingly, desire takes on more subtle forms for the psyche, but its importance remains profound; its impulse underlies all endeavour and is as essential and innate to the development of the human psyche as it is to the growth of any physical organism.

The Parasympathetic

I used the image of a tree at an earlier point, in the context of discussing values and I'll employ it again here in a different way; this time to help illustrate the relationship between, appetite/ desire, and action, with a simple analogy. While the roots of a tree reach into the ground, drawing nourishment from the contents of the soil, its branches and leaves reach out to generate new growth, using that material, by harnessing the energy of the sun. Neither the level of the 'roots' nor that of the 'leaves' can of themselves give structure or balance to the way the organism uses raw material and energy. The 'soil' supplies the raw material necessary for new construction, which largely results from the disintegration of other organic matter, while chemistry at the level of the leaves provides the energy for the action involved in its reorganisation - but the integration of the raw material into the metabolism of the organism takes place at what might be considered its central region of balance. For the tree, this is not so much the roots or the branches, but the trunk, from which the roots, branches, stems and leaves emanate and diverge, and their overall balance is sustained. For all organisms, at the level of biology this integration is directed by the DNA, which bears and expresses the established condition of equilibrium at their core. At higher levels of their organisation though, it is represented by relatively gross features of the organism in question. For the human being, the image of 'the trunk' in this analogy represents the central channel of the autonomic system, the *parasympathetic*, and it is here that the relationship between the sympathetic channels - i.e., between the spheres of emotion and action - is integrated and the patterns for 'new growth' are generated. In a sense, its contribution is passive as it doesn't engage directly in relations with the environment, nevertheless it provides and mediates the sustenance and wisdom of evolutionary experience that the spheres of the sympathetic draw on in their activity.

Evolution, in all contexts, necessarily proceeds through engagement and transaction within the environment. In the domain of human experience, the *sympathetic* aspect of the autonomic system is the principal instrument and mediator of that engagement. Through its left and right channels, we influence and are influenced by that environment, viz the term 'sympathetic', which suggests 'resonant'. These are the instruments of our experience, expression, and creativity. They have evolved integrated and inseparable, with the full weight and authority of evolution over hundreds of millions of years. Through their instrumentation we forge the shape of the future right now in the present, personally and collectively, so it's necessary for them to work together in a harmonious way in our everyday lives. It's also useful to be able to recognise some of the dimensions of their presence in spheres both personal and cultural. Meanwhile, the *parasympathetic*, the central channel of the autonomic system, constantly seeks to restore our organism to a state of equilibrium that reflects the continuous history of our evolution to date, and coordinates its continuing development based on that equilibrium. The purposes of the autonomic nervous system could be said to parallel and reflect those of our biological system, expanding the principles of metabolism into the sphere of conscious experience. By this I mean, the acquisition, digestion, absorption, and assimilation of ongoing experience (and 'reproduction' within the sphere of mental creation), with all the immediate effects for the organism, and the cascade of evolutionary consequences.

Reflecting its evolutionary roots then, the left sympathetic is associated with emotional experience and memory - in very general terms, the past - while the right side is associated with action and rationality, in anticipation and planning for the future, immediate, short and long term. These qualities are inseparably entwined and function together in generating the fabric of our everyday experience, but they are also highly adaptable and flexible. Accordingly, a combination of genetic inheritance and environmental influence may predispose the individual to identify to a greater degree towards the emotional or rational spheres, and to generate over time an outlook and lifestyle that reflects and reinforces that disposition. Such tendencies also become characteristic of collective behaviour within particular societies. Thereby do varied cultural temperaments grow, with all kinds of different permutations, but with tendencies towards the predominance of the 'right' or 'left' sympathetic in the sphere of collective behaviour. This finds its broadest expression in patriarchal and matriarchal tendencies in culture, the matriarchal tendency expressing more the emotional sensitivities of the left sympathetic in the collective attention; the patriarchal expressing the organising, rationalistic, tendencies of the right. (This distinction can be seen reflected at the level of political attitude, orientation and ideals, as well as religious organisation and symbolism.)

Meanwhile, the parasympathetic constantly endeavours to maintain balance, providing the 'centre of gravity' at the heart of our integrity as an organism, and as societies - the trunk of the tree - carrying the cumulative 'experience and wisdom' of evolutionary history. Talking therefore of the autonomic nervous system, including the relevant areas of the brain, not just in terms of neurology/electrical wiring, but as the central instrumentation of cognition, and with evolutionary processes in mind rather than psychology in a more conventional sense, certain observations may be made.

In everyday normal experience, all the cognitive functions of the senses are modulated by the *sympathetic* system, while the *parasympathetic* provides core governance and stability, and acts to maintain the balance provided by underlying evolutionary integrity. In the course of this, all of the information of experience relating to the sympathetic channels is processed by the corresponding hemispheres of the brain, which generate what might be called 'summaries' of that information. These are experienced by our conscious attention in that summary form, on the one hand as emotional response and on the other as preparation for action (reasoning, analysis, planning etc.). These impressions form the two principal spheres of conscious attention. These two 'fields of experience' are mediated by the sympathetic system and are the basis of the formation of emotional conditioning on the one hand (superego), and of the sense of being the 'one who is in control' (ego), on the other. Normally, these two aspects of our awareness appear merged, generating the perception of integrated psychological '3-dimensionality'. (I should just add the observation that this integration isn't an illusion, these two fields of cognition evolved inseparably together. The one couldn't exist without the other.)

The perception of 'self' is intimately associated with that of 'identity'; accordingly, we generally experience our everyday 'sense of self' in terms reflective of circumstance. It could appear therefore that the spheres of attention governed by the sympathetics represent the totality of personal identity, and that consequently there is nothing more substantial than that to the meaning of 'self', but it should be remembered that these spheres represent only two of the three significant dimensions of our psyche. The third dimension is expressed through the parasympathetic - the 'trunk of the tree'. From the perspective of our conscious attention, the parasympathetic provides the base line, or centre of gravity you could say, ever present at the seat of consciousness, therefore, at the heart of 'identity'. Although the constant activity of the sympathetics tends to dominate our attention, the parasympathetic represents the source from which the sympathetics draw, and which defines their integration. Its quality provides the central sense of existence, or *being*. Its contribution to the field of consciousness reflects the deep history of our evolution, back to its roots in metabolism (which, incidentally, continues to be represented in the autonomic system in the form of the vagus nerve) and beyond, to primordial origins. The intensity of the demands on our attention made by the activity of the sympathetic system, means that this profound but passive field of consciousness appears dormant, but nevertheless is ever-present and bears the power, presence, and authority of our evolutionary history, bearing great influence on the development of both the individual psyche and the collective psyche of humanity. To some degree, in all cultures, we carry an intuitive recognition of this, and it is reflected in our art, music, mythologies, religions, and philosophies.*6

There is much anecdotal evidence (including my own experience) that this deeper sense of 'Self' can emerge from a usually background presence to be represented much more directly in the attention, unobscured by the activity of the sympathetics, and yet remain essentially passive, with an enhanced 'witness' status you could say. Although unparticipating in engagement with circumstance, which is the territory of the sympathetics, the deeper Self is not without influence. This isn't direct however, but rather, a reflection of unbroken patterns of natural law established by evolution, since the beginning of time, woven into the fabric of consciousness. An awakening to this deeper sense of Self is implicit in the term 'Self-realisation', which is the subject of the next chapter. Observing the relationship of these three principal aspects of the autonomic nervous system offers some practical insight into the dynamics of the psyche in terms that relate psychology, physiology and 'spirituality' to their evolutionary background.

Looking outwards, everything appears to be composed of

'physical stuff', looking inwards everything is composed of 'consciousness', by virtue of which we experience the existence of all that 'stuff' - including ourselves! Which is the true basis of reality? Both really. Regardless of how we view the origins of consciousness, these also are two sides of one coin. Both directions of observation offer important understanding and insight into the nature of reality, but it must be said that, clearly, the status of the observer, which amounts to the nature of perception itself, is fundamental to what is observed and to how that knowledge is used. It is also the one aspect of experience that is within our scope to adjust without necessarily having to make changes to our environment. Perhaps contrary to the general tendency in Western culture to seek for knowledge and understanding by focussing entirely on external, objective, study, important advances can also be made by paying more heed to the position and condition of the observer... As I have discussed, a large part of this involves paying close attention to the way in which we use language and concepts and the way in which we tend to project assumptions into our descriptions of the world as established 'facts'; but another, at least as important, requires learning how to clear the faculties of perception from obstruction and clutter accumulated through everyday living experience. The essential ingredients of this second factor are already part of our innate makeup, built into the normal functioning of the autonomic system; complications arise from the excessive activity of the sympathetic system, simplification just requires allowing the sphere of the central channel, the parasympathetic, 'time and space' to restore balance.

However, while simplification requires reducing the identifications of the attention in its involvement with circumstance, it also requires the resolution of problems. The imperative of evolution means that circumstances must be engaged with, not ignored. Clearly, this is more manageable when mental faculties are in balance and working well. This requires a good degree of 'being in the present' rather than being tied to the past or fixated on the future.

This chapter and all the preceding ones include a fair amount of speculation across a range of ideas. Some of these are rooted in current scientific thought, some have roots in Eastern philosophy, some are just the product of my own creative inquiry, imagination and attempts to reconcile apparent differences or contradictions between those different ideas. However inquiry into the nature of reality and existence is framed, it seems to me that there will always be questions that logic and reason alone don't equip us to answer entirely.

Summary

Much of the behaviour and psychology of human beings concerns self-awareness. This is often conceived of in terms rather nebulous - philosophic or metaphysical, and its evolutionary background is poorly recognised or understood generally. Some of the practical context can be understood though, in terms of the functioning of the autonomic nervous system, which can help to make its form and significance clearer. Although our existence as an organism is physical, its form depends to a large degree on mental functioning, and this system is the instrument that mediates these two aspects of our existence. Viewing our behaviour from this perspective offers a practical approach that helps to place human behaviour, including the phenomenon of self-awareness, within the context of evolution, particularly animate evolution. At the same time, perceiving the pattern of the structure of this system draws attention to the profound depth of the origins and expressions of consciousness throughout living nature which have given rise to its emergence.

Chapter Notes

*1 the example of the nematode worm nevertheless represents a relatively advanced stage of nervous system development, i.e., the bilateral structure shared by most higher evolved life forms. This form of structure has been in existence since the early Cambrian era, but it is emergent from more elementary levels of electrical gradient communication within and between cells. Many other forms have also emerged, simpler perhaps, but very successful in their own way.

*2 Sri Mataji Nirmala Devi Srivastava, who was from India taught a very direct form of spiritual awakening called Sahaja Yoga. Here, the term 'yoga' means the reconnecting, or 'yoking' of the sense of one's individual existence with that of universal being, while the term 'sahaja' means spontaneous, indicating that like everything else in nature, this, owes nothing to conscious effort or achievement, but arises effortlessly from a deep desire inherent in the human psyche. This 'desire' is understood to be universal to all life and is central in the long standing Eastern philosophical view of evolution as what might be called a 'spiritual' phenomenon rather than purely 'material'.

*3 most cultures create personalised myths of one kind or another to reflect the perceived relationship of humankind to the universe. While these may help to illuminate 'divine' patterns in human nature, the opposite also often happens, where human behaviour is taken to represent universal qualities, projecting images of unbalanced human behaviour onto our view of universal nature; if we do it, this must be how the gods behave also...

*4 *known as 'virata', or great primordial being.*

*5 *although the term 'fields' is commonly used in relation to natural forces of a physical kind, it is still a concept of convenience, a metaphor. I'm drawing on that metaphor here, but it shouldn't be thought of in terms of 'physical' influence. The form of influence suggested here relates entirely to the domain of cognition.*

*6 *this sphere of consciousness has been referred to, again by Carl Jung, as 'the collective unconscious', a term which is easily misunderstood. It suggests, not so much the absence of consciousness, rather, a field of universal consciousness underlying all particular conscious experience, from which all such forms draw. I should just add that there exists a great variety of interpretation of the meanings of terms relating to the make-up of the human psyche. The interpretations I offer are based largely on my own understanding of these, from the perspective of evolution and the nervous system.*

Chapter 11

Self and Self-realisation

"The unexamined life is not worth living."
(Socrates 470 - 399 BC)

Outline

Although at an earlier point in this discussion I have made a point of distinguishing consciousness from self-awareness, with which it is often misidentified, nevertheless, our everyday experience of consciousness is largely represented, or you might say, encapsulated, in the form of self-awareness. The 'sense of self' has come to be a central feature in the expression of consciousness and is likely to remain central for a long time to come, as the human psyche continues to develop.

A distinction may be made, and has been in many cultures, between the perception of 'self' as a transitory reflection of circumstance, and a more profound sense of 'Self' that reflects a continuity of conscious experience in nature, with roots deep in the ancestry of life. This chapter considers further the meaning of 'self' and its place in the evolutionary scheme of things, and 'Self-realisation', as a natural development of human nature which has been recognised and described in a variety of different ways, and, again, in many cultures, over at least the last three thousand years.

In the discussion of evolution I've been presenting so far, the train of thought refers, as far as I can make it, to insights and understandings that have been created through the very practical and down to earth methods of Western science. These methods represent an approach to building knowledge on the firmest foundations evidence can provide - physically tangible, and testable. Not all natural phenomena are directly accessible to inquiry on these terms though, conscious experience being a significant case in point. But clearly, consciousness isn't a trivial phenomenon, it has a major, now dominating, influence on the direction evolution takes, and it merits a priority place in considerations. It can be difficult to do this though, without edging into territory that may seem more philosophical than practical.

Some readers may be beginning to wonder why I spend so much time talking about 'self', 'self-awareness' and so on. The reason is this; while most of our attention is dominated by the circumstances we experience, 'self' stands for *the experiencer,* not the objects/contents of experience, and the experiencer is clearly a fundamental aspect of reality, not to be overlooked or simply assumed without need for further investigation. From the perspective of evolution, clearly, both aspects of experience - the experiencer and the experienced - offer information and insight into the history of the universe and directions it may take in the future. Over the last few centuries, in furthering the understanding and control of the world of 'objects of experience', the attention of Western culture has focussed heavily on the physical side of the relationship between these two sides of reality. This has been of great benefit to us physically - and to a degree, philosophically, in raising the priority of reasoning over imagination and myth. However, when you think about it, one's own personal existence also reflects the entire history of the universe, from its origins, whatever they may be, through all the stages and domains of evolution, to this point, now, where we can consider and reflect on those origins. There is nothing new about this observation; in one way or another, most religions and philosophies revolve around the nature of 'self' and our relationship with the universe, indeed, some focus specifically on this inquiry, before and beyond any form of description or theology. The question, 'who am I?' has an ancient and honourable place at the heart of human inquiry, and I have little to offer that is original, what I want to do though, is consider the experience and place of self-existence within the terms and frame of evolution. This is where the physical and metaphysical meet.

The term 'self' derives from the sense of one's own existence as an entity, 'I am aware of my own awareness, and so, 'I' exist, as an object of my faculties of perception'. In a sense then, 'self' is an illusion of perception, and these two terms, 'self' and 'self-awareness' are inseparable concepts, yet, this perception is no more or less an illusion than that of any other object as a self-contained entity. In this sense, the essence of 'self' isn't the illusion, but rather, the conceptualisation of 'it' as an object. You could say that the experience of existence, our own or that of anything else, is the pure foundation of reality, never any descriptions we may make. In these terms, the inquiry becomes, 'what is the *essence* of 'self'?'

No 'object' truly exists independently of its surroundings at any scale, rather, the perception of its existence as a self-contained 'entity' is the meaningful factor here. At every point and level of observation, any entity we perceive as such, be it atom, grain of sand, human being, galaxy, or anything in between, is a coherent expression of natural elements, forces, and principles. This can be said to be the meaning of 'an object', the illusory aspect concerns the sense of separateness that 'object' perception tends to generate as a side-effect. There are cultural dimensions to the depth of this side-effect; the more a culture is immersed in the mentality of object perception, the stronger becomes this illusion. This is by no means a 'mystical' observation, it is completely, if somewhat paradoxically, supported by present day inquiries into the field of physics through which we have come to understand a great deal about the complete interconnectedness of all known phenomena through disassembling and examining in detail their components principles of organisation. Full appreciation of this interconnectedness though, requires standing back from the detail sufficiently to 'remember' the starting point of observation - undividedness. This can be said to apply at all scales, from that of subatomic 'particles' to human beings. How all this relates to the idea of 'self' is that if there can be said to be some fundamental principle, or quality, underlying the sense of 'self', it is simply conscious experience. The essence of 'self' is one and the same as the essence of consciousness.

For the majority of human beings self-awareness appears to be a basic feature of our mental condition. It used to be thought that this feature of awareness was unique to human beings. This was an intrinsic part of the Western view for many centuries, in which the human being was thought of as distinct from 'animals', before evolution became understood to be the basis of the common relationship of all life forms. In those terms the idea of 'self' is very close to that of 'soul'. But over the last century it has become increasingly clear that many other species have some degree of self-awareness; the sense of 'self' hasn't just appeared out of the blue as a feature entirely unique to the field of human awareness. It appears to be an emergent function of conscious experience of a simpler form - as an organism becomes capable of perceiving and manipulating mental 'objects' it begins to perceive its own existence (though this should be understood to be as an object of observation within the terms of its own dimensions of perception, not those of human mentality). For the most part, this 'sense of self-existence' may appear to be largely incidental and of little practical consequence, but in the case of human beings, it has grown to the extent that rather than just being something of a side-effect of perception, it has become a major factor in how we interpret and engage with our world of experience and therefore in the dynamics of evolutionary developments.

Reason and logic are great tools of inquiry and under-standing, but one of their effects is to draw observation down to a kind of neutral common denominator of functionality. When I think logically about self-awareness for example and try to analyse it, it tends to become an intellectual exercise, but in terms of actual experience it is really much more significant and personal; 'I am!', 'I exist!', 'I experience my own existence!' This isn't a phe-nomenon of reason or mechanics, and it isn't trivial. Self-aware-ness is at the heart and significance of our view of reality, and its essence reflects the entire history of the evolution of the universe! No small matter.

A certain amount of reasoning along the lines I've been presenting may help in recognising that the form the 'sense of self' takes is really a result of the contribution of many factors of perception working together. This could be compared to the structures of a telescope, or microscope, which use a number of different lenses, tubes, and so forth, to create their images, rather than just one 'lens' of perception. Similarly, we perceive one final view of reality perhaps, and a (generally) unified sense of 'self', but this perception is generated via many layers and filters. Some of these filters are detailed and particular factors of cognition, sensory experience, and conditioning that I've touched on, some are more general, concerning degrees of abstraction, and conceptual interpretation. All of them have evolved in intimate and necessary relation to the entire circumstances of our evolution over vast periods of time. Recognising, in principle if not in detail, the contributions made by these factors is relevant to understanding the outer qualities of self-awareness.

However, where the telescope is concerned, we are unlikely to lose sight of the fact that the central medium of its operation is light (i.e., electromagnetic radiation), but it's possible to pay so much attention to the constructs of perception that consciousness can become regarded as a secondary outcome rather than the central quality underlying all levels and degrees of cognition... The point of this analogy is, again, that the *core ingredient* of 'self' is consciousness itself, not any particular attributes of perception we may associate with 'it'. As I've discussed in Chapter 5, consciousness can be considered to have no 'shape' in itself. It can be thought of as more comparable to light than to the contents illuminated by light, while faculties of perception provide the various dimensions of experience through which awareness is directed and takes shape. Considering self-awareness against this background, naturally, my sense of self is shaped by the faculties of perception through which I experience my existence.

Although self-awareness is entirely abstract, i.e., non-physical, its powerful association with the environment means that it bears a very considerable influence on behaviour, individually and collectively, and has effectively become a field of attention within which we function, with 'real' consequences, material and psychological (and 'spiritual' you might say), and in which evolutionary principles are represented and expressed. Self-awareness can be viewed as being not just a curious expression of consciousness but a significant feature in ongoing evolutionary developments, having a role that seems, to me, to be a mental equivalent of the outer cell membrane. It isolates us from our surroundings, but at the same time, allows us to communicate and engage with them, as a kind of invisible 'bubble'.

Following the lines of reasoning I've been presenting, the sense of self may be considered an illusion of perception; but certainly, conscious experience exists, and if this exists, then the central ingredient of 'self' isn't illusory, we just make a mistaken description in exercising our objective faculty here. When we inquire into the nature of 'self' based on our experience of the world, we are investigating the reflection rather than the substance at the source of the image, which is conscious experience itself. When we turn our attention there, we find its origins deep in the ancestry of animate life, and prior to that, of organic life. We don't know where or when consciousness can be said to have originated, but it certainly didn't *begin* with the human mind. (Philosophical inquiry in India over several thousand years concludes that experience of existence is a primordial feature of reality, as fundamental as matter and energy, and is present as the essence of 'self' in all of the different expressions that may take. In the sphere of human experience, the gross, reflective 'self' is known as Jiva-atma; the greater, underlying individual 'Self' preceding reflection, as Atma, and the universal source 'Self' as 'Brahma', or Param-atma. Like light, these are, in essence, one and the same 'substance', you might say. In English, we use the word 'spirit' to mean essence, and this concept applies here also. 'Atma', in effect, means spirit, or essence.)

Metamorphosis

A number of traditions speak of a radical transformation in awareness that can take place for a human being, using terms such as enlightenment, nirvana, satori, being 'born again', moksha or mukta (liberation) and others. Because of many differences there are between these traditions, at first glance one may well assume that different experiences are being described which are unique to those cultures. Language and mythology generally lend an air of mystery and exclusiveness to such descriptions, but a closer look suggests that a particular, and not uncommon, human experience is represented here from the different perspectives of those cultures. If this is the case, is it possible to consider this in terms that can perhaps penetrate some of the more mythical language of presentation? As it happens, most cultures have drawn from observation of nature in their descriptions. I'm going to do the same thing here, but with more assistance from our growing understanding of evolutionary context rather than metaphysical or religious tradition. Here, I discuss this idea of transformation, as an innate feature of the psyche comprehensible to some degree in terms of the nervous system.[*1]

In addition to the ceaseless 'experimentation' of evolution, a phenomenon ubiquitous in the development of complex multi-cellular organisms, and widely recognised across many cultures is *metamorphosis*.[*2] Perhaps the best known and iconic example is the transformation of a caterpillar into a butterfly, but this is really a most dramatic and beautiful expression of a widely repre-sented 'technique' of nature, the origins of which lie in the

245

ancestry of the part of the tree of life to which most complex life forms belong. Another example is that of a bird; early development takes place inside the egg, then, as the chick develops it reaches a stage where it must break free from the shell for development to continue. Further examples that represent this principle are those of the acorn and the tree, and by extension, the relationship between all seeds and the adult organisms they produce - i.e., through the well-known phenomenon of germination. In all these cases, the encasing shell performs the function of protecting the fragile inner contents and containing their development until conditions are favourable for further development. Again, the requirement of isolation/protection is as ancient as life, first clearly represented by the outer cell membrane itself, common to all organisms.

More generally, the transition from childhood to adulthood also represents this principle of metamorphosis for many species, in a more subtle manner perhaps, but still bearing parallels with the transition of the caterpillar to adult form. The principle of transition of form which these examples represent, far from being rare and exceptional, is represented widely in complex life forms. Once recognised in fact, it becomes difficult to find complex organisms that *don't* express 'metamorphosis' in some way. Overall, metamorphosis represents the existence of innate patterns of development within an organism, which result in significant transformation of its form, in stages, sometimes rapid, sometimes gradual. But this doesn't apply only to physical features, whole patterns of cognition and behaviour are also involved in the case of any creature with a nervous system - sometimes involving the re-organisation of the nervous system itself. (Check out sea squirts. These are sea living, complex creatures that look a bit like sponges, but have a nervous system that belongs on the same evolutionary line as our own, albeit distantly. As these creatures develop from a larval stage, they find a suitable place to attach themselves and then absorb the navigational part of their own nervous system, i.e., their 'brain', because it is no longer required for their adult life as filter feeders. Beautiful creatures, by the way.)

There is a view, particularly in Eastern cultures, that similarly, the human psyche is 'complete', yet contains stages of development that generally remain dormant, but may awaken in the attention, given suitable conditions environmental and internal. In this view, all the physical aspects of the nervous system and brain may be developed and integrated, but a further stage of development takes place in which self-identification is spontaneously released from the restrictions imposed by the sympathetic system in the form of ego and conditioning (discussed in the last chapter), and the awakening of a deeper, more universal, perception of reality.

The image of an egg is often used to represent this transition within the psyche. An egg is recognised to be complete, but that completeness is also a prerequisite for its further development into a fully formed hatchling. Again, the hatchling is complete in its own right - and this is essential to its further development into an adult bird. In evolutionary terms you could say, 'completeness' doesn't represent a 'finished product', but rather, a continuous condition of development, which we may perceive as distinct stages. Clearly, objective evidence in support of this being a feature of the psyche is slim because here it concerns the nature of perception itself rather than any overt physical attributes, nevertheless, anecdotal evidence exists in the form of descriptions from many different cultures, in their own particular terms, over many centuries. Descriptions of this kind tend to concern the personal quality of experience rather than physical terms. More subtle it may be, but the quality of experience underlies the 'muscles and sinew of culture', and the psyche is the central instrument through which this is generated and expressed. Many natural principles of life are reflected in the sphere of human conscious experience, after all they are the basis of the evolution of the psyche. Observation suggests that metamorphosis is one of these principles.

Self-realisation

I'll begin this piece by saying that I have some personal experience of the topics I discuss here. That's not to suggest that I am 'expert' by any stretch of the imagination, just to make it clear that this isn't a purely intellectual exercise.

The term 'self-realisation' is one that has become increasingly used in Western culture in recent times, in ways that reflect particularly Western psychological ideas of self-development. Prior to that, it can be found used in philosophical writings of the Advaita Vedanta tradition of India, going back many centuries, where it is understood in quite a different way that is more relevant to the observations I want to make here. From a Western perspective the term 'self-realisation' generally suggests an idea of achievement of personal fulfilment and completion of identity developed through one's engagement with the environment and dedicated endeavour of some kind.*[3] From a more Eastern perspective however, Self-realisation refers to a very particular transition, or metamorphosis, from the self-awareness of general everyday experience, to the awakening of identity at a fundamental *inner* level, in a spontaneous act of recognition - or 'realisation' - of completeness, before and beyond any definitions of circumstance or description. An important result of this experience is a great reduction in the degree of 'identification of self' in terms reflective of the environment and circumstance. This has ongoing consequences, over time, in the realignment of behaviour, as personal discrimination of truth and value becomes trusted and exercised with greatly increased insight. Realisation in this sense is known as 'Self-realisation', where the capital 'S' indicates this deeper, more universal 'sense of Self' than is generally experienced, which is represented here throughout, by the expression 'self', with a small 's'.

Self-realisation in this more profound sense can be said to have two aspects; first, 'realisation', suggests recognition, and refers here to the direct awakening to awareness of the deeper fields of consciousness/being underlying the abstract, relative layers of thought. This, like all instances of recognition, is not cumulative or time dependant but a spontaneous event of awareness, and it may occur at any point in a person's life, without regard to time or conscious effort spent in its pursuit, or, for that matter, the absence of such effort - nevertheless, not arbitrarily. The source of the inspiration through which this shift in attention takes place is understood to be a largely unconscious desire for Self-knowledge, an innate evolutionary imperative, deeply rooted in the psyche, that could be compared to the germinating principle of a seed. Although this desire is generally dormant, or unconscious, it underlies a lot of conscious activity, finding expression in all kinds of aspiration, inspiration, inquiry, and creativity. When conditions are suitable, it may awaken directly in the attention, in the experience of Self-realisation. This recognition is a profound event in a person's life, often referred to as 'enlightenment', or 'rebirth'. Self-realisation isn't a mystical experience however, although it is often regarded as such, but rather, one of profound *normalisation*, as identity finds its deepest source - which was never truly absent, just obscured by the focus on circumstance, as the sun is obscured by clouds.... Once awakened, this recognition becomes more easily sustained, never forgotten completely, contributing to our wider understanding of the world, even though clouds continue to come and go as part of nature. Mystical no, but profoundly significant.

The second aspect of Self-realisation is reflected in another meaning of the term 'realise', which is, 'to make real'. This does involve time and experience, relating to the longer-term effects that Self-recognition has on one's more relative sense of identity, and is perhaps closer to the Western psychological view of self-realisation. To grasp this idea, although the attention is spontaneously awakened and expanded in the initial experience of recognition, or 'realisation', there are many established behaviours of habit and conditioning, gross and subtle, that don't - and can't - change completely overnight, but which begin, sometimes rapidly, sometimes more gradually, to realign to this new, more profound, perspective. In this sense it can be said that awakening to the deeper, more universal dimension of our awareness, leads to the spontaneous realignment of our everyday identity towards correspondingly more universal values, but no amount of pre-conceived attempts at such realignment, adjustment of lifestyle, philosophical inquiry, intellectual analysis, or religious teaching, can lead to that awakening of itself.

This is not to say that awareness of such a development is therefore pointless. On the contrary it seems important that society should have knowledge that such a significant evolutionary principle is fully active at its heart, within the individual, expressed in terms that are appropriate for the age in which we now live, and a recognition that our approaches to social organisation can support or hinder its emergence. Again, this may appear contradictory at first glance; if Self-realisation is an innate feature of the development of the psyche that frees the attention from identification with circumstance, how can adjusting circumstances contribute to its expression? This comes back, again, to the point made earlier, about the relationship between nature and nurture. Briefly, a plant grows according to its own unique, innate, qualities, but it requires wholesome supportive conditions in the environment for those qualities to be fully expressed. This also applies to the full expression of the potential of the human psyche. Conditions in the environment may support or obstruct that development. It could be said that most cultures instinctively recognise this, to varying degrees, and seek to create conditions that are supportive. However, trying to impose conditions for growth, through religious doctrine, or political ideology, for example, tends to implicitly place discrimination and insight under the shadow of conditioning and dogma. There can be a fine balance between the life enhancing support such structures can provide and the restrictions they place on growth.

There is nothing new about the phenomenon of Self-realisation. It has been a normal part of human existence for many thousands of years, possibly from the earliest times of the emergence of human species and has been described and documented by different cultures in terms that suit them best, particularly by established religions but also out-with any formal bodies. Examples of such description are contained in the teachings of Christianity, Islam, Buddhism, Hinduism, Sikhism, Zoroastrianism, Judaism, and others. However, no religion or philosophy has a monopoly on this knowledge; it can be, and is, described in many ways, but description can never provide access to this area of experience on its own - nor can any amount of intellectual analysis. However it may be described, this appears to be a feature of human development that occurs spontaneously, regardless of religious belief - or its absence for that matter - and with or without description. Essentially, it's a living development within the sphere of consciousness. Perhaps paradoxically, the fact that this transition is entirely within the province of nature means that it has generally been described in somewhat mystical terms. It appears to be a significant feature of our existence but beyond manipulation, and so, viewed as 'mysterious' (although religious establishments of different cultures generally view themselves as sole facilitators of this transition, mediated through priesthood in one form or another).

Although Self-realisation is a profoundly personal experience, it can be seen that there is an evolutionary context to its emergence as a cultural phenomenon. In many very different and geographically distinct societies, from the indigenous cultures of Australia and North America, to those of India, China and the Middle East, traditions have developed which reflect a perception of the deep roots of the human psyche in nature. In these cultures, behaviour and practices that encourage sustenance of this awareness as part of normal everyday life have been created, recognising its precedence to man-made conceptual rules. At least in principle, the overall purpose of such behaviours is the cultivation

252

of an open, balanced state of mind that remains sensitive to the knowledge and wisdom of nature.

In the course of time however, many of these traditions have become institutionally ritualised and mythologised by established religions, or otherwise degraded, often becoming significant obstacles to awakening rather than aids to their removal. Ultimately, it can be said that Self-realisation is within the provenance of nature to gift, or not, as with everything else natural, and has nothing to do with belief, ritual or style of dress... It's also understood that this 'awakening' spreads collectively, not through words, ideology, or teachings in any formal sense though, but rather in much the same way that people engage and interact generally, person to person - in a form of mind/mind communication and recognition.

In part, this can be understood in terms of evolutionary imperatives that underlie developments in the domain of culture. For some people at least, the desire for insight intensifies as living conditions grow ever more complex, and as contradictions and conflicts inherent within many long-established views become apparent and hard to reconcile. This affects all fields of inquiry and can be said to represent an 'appetite for truth'. From the arts for example, to the sciences, or through the constant demands of social justice, it finds expression in all different spheres, sometimes subtly, sometimes more forcefully. Combined with a tendency for insight to be converted to 'belief'; as-it-were, stepping stones of knowledge, it can and does lead to many 'fragments' of partial truth in all of these areas. Such partial truths are important and necessary features of practical life but don't contribute directly to a holistic sense of the 'meaning of life', often a greater sense of fragmentation. From that perspective they present more like pieces of an unassembled jigsaw, Self-realisation, however, represents this appetite finding satisfaction in a more central way, through the direct awakening of a sense of integration, preceding and underlying detail of all kinds. On these terms, this isn't just a matter that concerns the individual but is an important feature in the evolution of society that may be finding spontaneous emergence on a greater scale than ever before, a flowering of insight.

Some readers might find this idea rather optimistic, considering the state of the world; there is a lot of chaos, inequality and injustice at all sorts of levels, with no great sign of that situation improving much any time soon. But it seems clear, looking back on history, that many aspects of human society have improved a great deal over the course of the last few hundred years, most rapidly since the end of the second world war. There are many reasons for this, practical ones such as rapid developments in technology and global communications, physically and remotely, have contributed significantly to expanding and raising the awareness, opportunity and expectations of people, in more affluent societies particularly, and in general globally. But the calamities and warnings of the 20th century have also brought awareness of the need for international cooperation to an unprecedented level, considering the great problems that remain. The perception of chaos is, naturally, more attention-grabbing than developments in harmony and cooperation because of the sense of threat and insecurity it generates, but the counterweight to this is the appetite for harmony it inspires, and pressure towards truth is a powerful element of that desire. Awareness of problems is a first requirement towards dealing with them, and discomfort is an essential ingredient of that awareness.

Another very basic factor is this; society is composed of individuals, and the quality of personal behaviour contributes to the quality of collective life in every area. From home and the relationships between parents and children, to the sciences and the arts, politics and everywhere in between, society is nourished and strengthened by the presence within it of sensitivity to deeper currents of evolutionary integrity and impulse. This, not only at the level of specific insights, ideals, or concepts, although these contribute, but at a still more fundamental level, of balance and perception in every area. Again, there is nothing original about this observation of course, the entire notion of personal virtue reflects a recognition that collective values have their roots in personal sensitivities. This relationship between the individual and the society they inhabit has effects at every level of culture, from the scale of day-to-day personal interactions to that of global activities. In all matters, personal discrimination is a crucial ingredient and underpins the realm of social values. Self-realisation makes an important contribution to these processes, allowing perception of circumstance to be innately more sensitive to evolutionary imperatives. This, from a level of integration greater than emotional response, rationalisation, or dogma can provide. Conventionally, this wider sense of context and consequences is assumed to be provided by cumulative experience, individual and collective. We would usually regard this as the basis of wisdom perhaps, but where Self-realisation is concerned, the idea of wisdom relates more to the innate and spontaneous 'wisdom' of nature, expressed *now*, through discrimination in the present, and ever-present at the core of cumulative wisdom.

The desire to know and understand constantly strives to reach beyond limitations. The nature and function of conceptual thought is however, to create new constructs to which we get attached... One of the lessons of the East, reiterated in different ways, is that it is also a normal human capability to outgrow identification with concepts, to address and satisfy the deeper appetite of inquiry by reconnecting with nature - and therefore the meaning of life - at a more fundamental level already present in our psyche. This is the relevance of Self-realisation, both for our quality of life, and as a source of insight into all the deeper questions we have.

Self-realisation as a Collective Phenomenon

Self-realisation is usually thought of in terms of fulfilment of the individual, because the concept 'self' is, of course, strongly associated with the inner, personal, sense of 'self' existence. Viewed in terms of evolution however, it can be said that its broader significance lies in the collective sphere because this is where the dynamics of all evolutionary development take place, and to which this awakening contributes. The individual person provides the basic material for developments at the collective level, and accordingly, the quality of their contribution is important. Each of these spheres, the individual and the collective, is equally essential for further development. At all levels of nature this remains the case, but the actual dynamics of complex development take place in the collective sphere, rooted upon, but beyond the scope of, individual beings. As an individual our primary interest necessarily lies in our own personal condition, physically emotionally, and spiritually, and it can be difficult to see beyond that perception of 'self', but our personal satisfaction is part and parcel of the greater, collective dynamic that is ongoing and, personal interest not-withstanding, is not the priority here.

Whether we consider molecules, microbes, ants or human beings it is clear that coherent collective behaviour is capable of much greater feats than individuals alone can achieve, though their contribution is fundamental. The same thing applies to the phenomenon of Self-realisation, and its significance lies as much in its contribution towards the ongoing developments of the human psyche as a whole as in its benefit to the individual person.

Summary

Self-awareness is a medium in which we function and engage, like the water in which fish swim, inclusive and essential, but invisible to us on account of its all-pervasiveness. Generally, reflections of the environment dominate in the perception of 'self', but the deeper awareness at its heart is much more profound, rooted in our entire evolutionary heritage. Most cultures have come to recognise and describe in their own terms, that the transient surface of self-aware-ness can give way spontaneously to reveal this more profound basis. The principle of this 'awakening', represented in the term employed here, Self-realisation, can be understood as one of metamorphosis, which is widely expressed and represented in nature, not least in the process of germination, which takes place when conditions are fa-vourable. Environmental conditions are important, but the essential nature of this transition is inherent within the structure of the seed. The perspective of evolution offers a way of understanding and de-mystifying this principle as an important feature of human develop-ment, to at least some degree.

Self-realisation, viewed in these terms, relates to the emergence of the attention from a stage of development under the confinement of the ego and superego, into a further stage where, although these aspects of the psyche continue to exist - as necessary to our practical existence as arms and legs - the nature of their functioning becomes more that of creative freedom than restriction, of wings rather than shells. As far as evolution is concerned, you might say, the relevance of this is at the collective level just as much as at the personal.

Chapter Notes

*1 this idea raises the observation that metamorphosis can only take place where the advanced condition has already been established in the species concerned, it can't just suddenly appear 'out of the blue'.

*2 the term metamorphosis is used in scientific contexts, with a particular meaning. My use of it here is in a broader, colloquial sense.

*3 this view is put forward in slightly different ways but has its origins in psychological theories of the last century or so. Principle among these was that of Carl Jung who offered the concept of 'individuation', which encompassed the idea that the individual is on a personal journey of self-discovery, through which he/she can develop their full potential, not materially but psychologically, which is unique for each person. This essentially represents self-realisation as a 'process of becoming' - of self-fulfilment and discovery through experience, over time. There are other variants of this view, such as self-actualisation for instance, but Jung's view is probably closest to the Eastern view as it represents the process as being an innate feature of the human psyche, and therefore essentially spontaneous in nature (in addition, Jung was well read on matters of Eastern inquiry). I don't intend to explore those variants here, I just want to point out the distinction between the view that such a principle is an innate feature of human nature and the view that it is one that is 'attained' through effort of will of some kind, physical or mental.

Chapter 12

Dirty Hands

One might anticipate that the last piece in a series of discussions of evolution would be about possible future developments. In a sense this piece does concern that, but it isn't an attempt at prediction by any means, rather, it concerns the reality that we are currently in the process of creating the future, for ourselves and humanity, and for the direction evolution is taking on Earth. I am tempted to present a long-winded take on the responsibility we have towards the environment, the planet, climate, humanity and so on, but for the sake of the reader, and my own, I'll resist that, in the assumption that those who have had enough interest and patience to read this far are likely to be well aware already of these matters to some degree. The central point I want to make here is much simpler than that. It's my belief that we view and interpret the world around us according to our own mind-set, and we always have some freedom of choice in this. This isn't just a matter of positive thinking on my part, simple observation makes it clear; and the most important evolutionary domain we occupy now is no longer that of genes alone, any more than that of molecules, but of mind. Its time scale isn't millions of years, it is taking place right now, and the 'mind' concerned is our own.

Earlier in this work I've made the point that the individual mind is the seat of discrimination of truth, interpreter of reality. This is the central and most important faculty we have, and we use it in relation to everything we experience, externally and internally; that is, about the world that surrounds us, and the world of our own thoughts, what we think, what we 'believe'. With this faculty we shape the world, including ourselves, and here lie our greatest restrictions and, potentially, greatest freedom.

Based on rationality alone we *could* choose to regard evolution as blind, random, meaningless, directionless, mechanically predetermined and our own existence as a reflection of that, pointless, purposeless and irrelevant - we are meaningless observers of a meaningless universe. Or we may choose to gauge our measure of evolution, in all its domains of expression, in terms of its manifest harmony, balance, and diversity, and to recognise that our own existence, as 'experiencer', is a product of that - and at the frontline of developments. We occupy a very special and important position, as a major fruit of evolution, but we are by no means its only fruit. On account of our position we are, in effect, custodians and representatives of nature, and our freedom to perceive, conceive and choose our place within it and the actions we take, is an essential feature of that. My own personal insight is that the future of human existence and the shape of life on earth is one of possibilities not predetermined certainties, whether conceived of as 'divine' or 'blindly physical', and these we choose and create through our perceptions and actions - and failures to act.

Personally, I trust in evolution, in the terms I have tried to outline; being its product I don't really see how I could do otherwise without doubting the validity of my own existence, and of my opinion in the matter... In the end, everything I believe and understand is rooted in my own ability to distinguish what is true and relevant. To do that requires attending to some degree to maintaining conditions for clear, unbiased, perception and thought. For those who may consider that I am getting too philosophical here, yet again I return to an earlier analogy of the craftsman's tools. I don't see anyone taking issue with the idea of a carpenter preparing and maintaining his tools for use; the preparation of tools is essential to the tasks they are required to perform. This is by no means a modern idea, but one which has been clearly expressed by different cultures over many centuries - in order to know about the world around us, particularly in these complex times, the first requirement is to have some knowledge of ourselves.

Our minds are the result of around thirteen and a half billion years of universal evolution, since the 'Big Bang' (whatever that was... It could be said that the 'Big Bang' concept represents the point at which our capacity to understand the world through reason begins, leaving behind a universe unframeable, unquestionable and unanswerable). We should trust these minds, value them and use them well, but keep, as far as possible, attentive to their strengths and limitations. Herein lies our importance, and our profound meaning as instruments of evolution. By this I don't mean that everyone should sit around trying to understand everything. In practice, it's only necessary to see as clearly as we can what is in front of us in our own circumstances, and our mind does this best if allowed to do its work with a minimum of obstruction. Most obstruction originates in the social environment but becomes internalised, in the form of conditioned attitudes and adopted views of the world. In my experience, the mind is naturally capable of seeing through and beyond these restrictions, given the opportunity and inclination.

But I'd like to make another, related, observation. Most

religions and philosophies emphasise the quality of the individual and their spiritual/psychological wellbeing, which seems perfectly reasonable, but I no longer believe that personal spiritual emancipation *alone* represents the priority 'purpose' of evolution. This may have been a more useful view in earlier times when many less people populated the planet, but it is clear that the scale and nature of the collective behaviour of human culture now has a major impact on the quality of life on Earth as a whole, for all the populations who must share it, for our children and their children. In all dimensions of our activities this is now a matter of urgency. My own view is that we shouldn't stand aside from the complex and diverse challenges of society or the world at large out of distaste for argument, or fear of being 'contaminated' by the effects of complexity, but should engage with them and address them directly as best we can. In the past, perhaps it was more relevant and best for those who could, to focus on personal emancipation and let the rest of society go its own way, but the problems now are global and pressing. Personal dimensions and conditions can't be detached entirely from the responsibility to participate in the practicalities of the world, quite the opposite. This is how change happens. This is how we evolve collectively. The future of human society and the direction of evolution depend on it. In the end, one way or another our actions make the world. Not by wishing or hoping - or praying, or fate or magic, mysticism or the alignment of planets, or leaving it entirely to others to deal with, but through our actions the world is carved, by accident or intention - mostly a rich mixture of both. If there should be such a thing as providence, divine or otherwise, even so, we are now the hands of evolution. We're participants, there's important work to get on with, we can't be afraid to use these hands - even if that means getting them dirty in the process.

Postscript

A few final comments.

In this work I have tried to introduce the reader to some ideas about evolution, many of which are well known and established, some much less conventional, a lot just my own observations. I have skimmed over all of these because I wanted to give an overview rather than immersion in detail. Details are important too of course, and there is much of interest to explore, but for the sake of clarity I've deliberately concentrated on the 'bare bones' of the ideas involved, which seems somehow appropriate.

To give a little more personal context, at the time of writing this I am in my seventies and I have been thinking about these matters since I was 20, so, although I have no formal learning and am definitely not mainstream, I haven't just jumped into these trains of thought as a passing fancy, I've been travelling and exploring with them for the greater part of a lifetime. When I began writing however, I hadn't intended to discuss evolution in the broad sense that it has turned into, my main interest initially was in 'social values', i.e., what people consider to be important as principles of behaviour, individually and collectively. It was only as I began to explore that subject and to recognise how profoundly integral 'values' appear to be to the developments and nature of culture that I found myself more and more considering their evolutionary roots. Nevertheless, the subject of 'values' retains a place at the heart of these considerations.

The development, or evolution, of insight, isn't restricted to purely practical matters of course, but takes place in all dimensions of human value experience. All significant cultural developments, from the sciences to great religious and philosophical traditions, to political ideals, are invariably associated with individuals who may be accordingly revered in their way, but the impulses and aspirations which they reflect originate, appear and resonate within the collective attention. In any sphere - science, social values, or spiritual insight - once such insights are awakened and focussed, they become part of the living mortar of culture, upheld (or eliminated) by the power of nature, and 'naturally selected' for the ability to support and sustain further development. Such insights belong to no-one and everyone.

The astute reader may have noticed that questions of the existence or otherwise of 'God' don't figure in this book... This is chiefly because I find that this kind of discussion can short circuit inquiry into the nature of existence by leading into a diversion of thinking in terms of 'creationism versus atheism', which is about as useful to inquiry as political argument... That's not to say that there is no useful conversation to be had in this regard, just that it isn't relevant here. I'll just say that it seems to me that regardless of the understanding and insights that can be drawn about the processes of nature, the essence of existence - most especially, one's own existence - can't be explained in terms of 'how' or 'why'. With a net you may attempt to catch and describe everything that water holds, while giving little or no attention to the water itself... Consideration of the 'mechanics' of evolution by no means reduces the sense of awe about existence, nor can it deal with big questions of heart and mind that can't be expressed or answered in concepts or mathematical formulae.

Finally, the ideas I've presented here derive from a combination of my life-long interest in evolution in the Western sense and further insights gained through the support and teaching of Sri Mataji Nirmala Devi Srivastava to whom this work is dedicated.

To be continued...

Reviews

The book, Universal Evolution, contains a thoughtful philosophical approach to the origins of matter, life and consciousness. With care, the author leads the reader from subatomic domains to those of higher complexity thereby offering a top-down thesis by which he explains the similarity in the evolutionary processes leading to the atom, the molecule, organisms and finally the Self. Ultimately, he describes Self-realisation as an integral feature of human evolution.

Daniel Fels PhD

(Evolutionary Biologist and Specialist in Electromagnetic Cell Communication)

This is an amazing piece of work that takes on some of the most profound questions on a range of matters and explains them in a way that those who are not familiar with some of the subjects involved can follow and think about deeply. The author has taken great attention to explain some profound questions in a manner approachable for those who get bamboozled with scientific theory etc. It still has plenty that may bamboozle, but it opens up plenty fresh insights, and raises as many questions as answers for a reader like me. It's a real page turner for those who are interested in this type of subject.

John Lehmann (Musician)

I was asked to review this book by Lovereading. Wow what a read and will always be apt. At present with conflicts around the world, I keep asking the questions why are humans the way they are? How are we shaped? How did evolution occur? Reading this book is less complicated and the author gets it. I had not thought about evolution from a process point of view as spiritualism sometimes overtakes and we then look at this really from just one perspective. Here the author looks at this from all angles. Some are known as we have grown up with Darwin's theory but it is also interesting from the author's perspective also. It is all part of the bigger picture. Well researched, well written and author's exploration of his own theories also.

Jane Brown (LoveReading Reviews)

www.ingramcontent.com/pod-product-compliance
Lightning Source LLC
Chambersburg PA
CBHW031041110426
42740CB00046B/286